SpringerBriefs in Microbiology

For further volumes:
http://www.springer.com/series/8911

S. N. Chatterjee · Keya Chaudhuri

Outer Membrane
Vesicles of Bacteria

 Springer

Prof. Dr. S. N. Chatterjee
Biophysics
Formerly of Saha Institute of Nuclear
  Physics
Kolkata
India

Prof. Dr. Keya Chaudhuri
Molecular and Human Genetics Division
Indian Institute of Chemical Biology
Kolkata
India

ISSN 2191-5385   ISSN 2191-5393 (electronic)
ISBN 978-3-642-30525-2   ISBN 978-3-642-30526-9 (eBook)
DOI 10.1007/978-3-642-30526-9
Springer Heidelberg New York Dordrecht London

Library of Congress Control Number: 2012941626

Printed on acid-free paper

Springer is part of Springer Science+Business Media (www.springer.com)

Preface

This monograph deals with the outer membrane vesicles or OMVs of bacteria. Although some isolated studies have recently reported the production of OMVs by a few Gram-positive bacteria and other organisms, these have not yet produced any significant impact *vis-à-vis* that produced by numerous studies on Gram-negative bacteria. Accordingly, we have decided to focus on the studies of Gram-negative bacteria in general in respect of the OMVs produced by them.

A systematic study carried by one of us (SNC) along with his collaborators revealed, for the first time, the details of the production of OMVs by a Gram-negative bacterium, *Vibrio cholerae*, in the years 1966 and 1967. The OMVs were found to be produced by the bulging out of the cell wall at places, production of a constricted neck of the bulged out portion and release of the vesicular structure so produced into the extracellular medium. Besides recording these findings, it was interpreted that this phenomenon represented a novel secretory activity of the bacterial cell. This finding was, however, challenged by a contemporary publication in that the OMVs were reported to be a product of cell lysis and of artefactual origin. On the other hand, some other contemporary studies by several investigators including the original reporters (SNC and collaborators) ruled out the artefactual origin of the OMVs and established that they are produced by the cells during their active growth phase. In this context, the present authors have thought it worthwhile and relevant to record in this monograph, before it is lost or become nonavailable to future researchers, a brief revisit to the early days of the discovery of OMVs, the challenge posed against their identity as a distinct, important and useful functional entity of the Gram-negative bacteria in general and its overruling by the subsequent contemporary investigators.

Since then, studies on OMVs of Gram-negative bacteria have been progressing steadily and have brought out not only their physical and chemical nature but also, and more importantly, their diverse functions which include (1) delivery of toxins, antibiotics, etc. to host cells, (2) transfer of proteins and genetic materials to host cells, (3) cell-to-cell transfer of signals through chemical molecules, (4) elimination of competing organisms, (5) promotion of bacterial survival and pathogenesis in the host, etc. Recent studies on the proteomic profiles and biogenesis of OMVs have been revealing many other aspects of their uses including development of diagnostic tools and vaccines and antibiotics against pathogenic bacteria. It is expected that future studies on OMVs will play a great role in the development and production of effective vaccines against pathogenic bacteria which are challenging at present the very survival of humans and animals in this world. This monograph has presented a brief account of the physiological and medical applications of OMVs and particularly their use as effective vaccine against the meningococcal diseases caused by *Neisseria meningitidis* in different countries of the world. This book has opened up the scope for doing further researches on the preparation and use of different types of OMVs, i.e. native OMVs, detergent treated OMVs, engineered recombinant OMVs, etc., and their suitability or rather efficacy as vaccines against various human and animal pathogens including *Vibrio cholerae* against which no suitable vaccine is yet available.

Ever since the discovery of OMVs in the early 1960s many illustrious investigators have contributed enormously towards the development of our knowledge of OMVs as very useful and important functional entities of Gram-negative bacteria, leading ultimately to their use as effective vaccine against the meningococcal diseases to start with. But no account of this story will ever be complete without the mention of the contributions of two prominent groups, one led by T. J. Beveridge of the University of Guelph, Guelph, Canada and the other by M. J. Kuehn of the Duke University Medical Centre, Durham, NC, USA. While the researches carried out by these two groups revealed many of the fundamental properties of OMVs and established the solid foundation for the subsequent researches on OMVs, the publication of several elegant reviews at different times by the group led by M. J. Kuehn has helped immensely in our understanding of the nature and function of OMVs and in giving indications of the direction of future researches required in this area. While this monograph has attempted to record faithfully their contributions including those of others, the authors beg to be excused for any unintentional omission.

We would like to express our sincere thanks and gratefulness to Springer-Verlag, Heidelberg, Germany for kindly agreeing to publish this monograph as Springer-Brief in Microbiology, to their publishing Editor, Dr. BrittaMüller for kindly giving us their acceptance news and providing us the copy of the Publishing Agreement for SpringerBriefs and to Dr. Andrea Schlitzberger, the Project Coordinator for kindly helping us in the preparation of the manuscript and promptly answering to our day-to-day questions. We are thankful to the members of the Molecular and Human Genetics Division, CSIR-Indian Institute of Chemical Biology, Kolkata, India and particularly to Mr. Avirup Dutta who performed the transformation of all the hand drawn figures into print-friendly digital form and to Ms. Debashree Chatterjee and Mr. Arun Roy for providing technical and other help. Both of us (SNC and KC) are grateful to our respective family members for their kind patience, support and all possible help during the period we were busy with this project.

Kolkata, India   S. N. Chatterjee
                                                                                             Keya Chaudhuri

Contents

Chapter 1
Discovery of the Outer Membrane Vesicles: Revisiting Contemporary Works

Abstract The discovery and mechanism of formation and release of outer membrane vesicles (OMVs) by the actively growing *Vibrio cholerae* cells and their interpretation as a novel secretion mechanism are revisited. Also a brief account of contemporary works, an initial criticism describing that the OMVs were of artifactual origin and the subsequent re-establishment of the fact that the OMVs are not a product of cell lysis and are formed during the active growth of Gram-negative bacteria, have also been revisited for the historical record.

Keyword OMVs · Discovery · Mode of formation · Criticism · Confirmation · Contemporary works

1.1 Introduction

The bacteria-free culture filtrate of *Vibrio cholerae* cells were shown to be enterotoxic by the rabbit ileal loop test (De 1959). This work presented, for the first time, the idea that the cholera toxin is an exotoxin and is secreted into the culture medium by actively growing bacteria. In order to get an idea of how the toxin molecules were being secreted, one of us (SNC) along with his co-workers started investigating, in the early 1960s, the ultrastructure of *Vibrio cholerae* cells harvested from different phases of growth. From contemporary knowledge of the size of the toxin molecule it was expected that the secretion of these molecules by the *V. cholerae* cells could be visualized by the resolution provided by the electron microscope.

S. N. Chatterjee and K. Chaudhuri, *Outer Membrane Vesicles of Bacteria*,
SpringerBriefs in Microbiology, DOI: 10.1007/978-3-642-30526-9_1,
© The Author(s) 2012

Fig. 1.1 Ultrathin section of *Vibrio cholerae* harvested from the logarithmic phase of growth in peptone water; stained with lead. Many cell wall bulges can be seen. The sequence of the bulging process is indicated by the sequence of letters A, B, and C. *Arrows* indicate the association of granules with the bulged-out cell wall membrane. Bar represents 0.1μm [From (Chatterjee and Das 1967)]

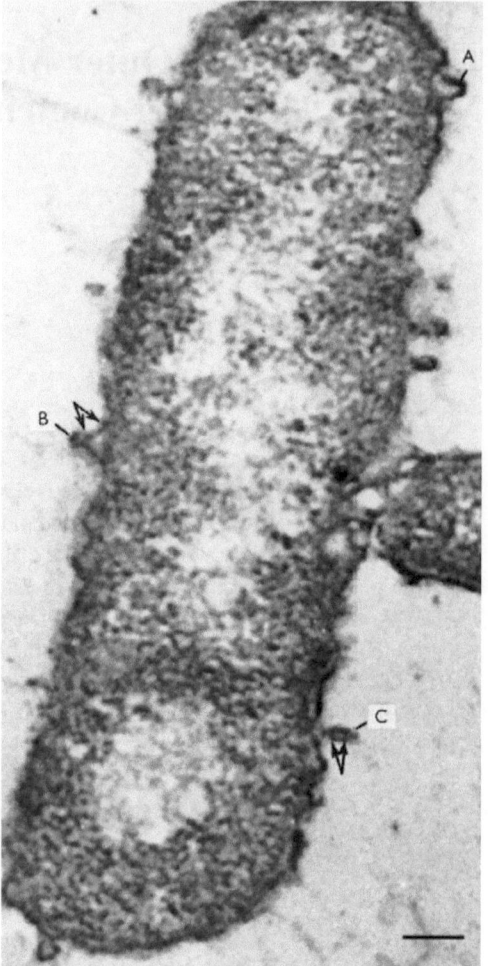

1.2 Discovery of OMVs

Study of the ultrastructure of *Vibrio cholerae* cells harvested from the logarithmic growth phase in a nutritionally rich liquid culture medium revealed bulging out of the cell wall at different localized areas (Fig. 1.1), while otherwise maintaining normal cell structure (Chatterjee and Das 1966). The bulged out portions of the cell wall were found to be pinched off and released into the extracellular medium as globular vesicles (Fig. 1.2). This phenomenon was considered as a novel secretory mechanism of *V. cholerae cells* (Chatterjee and Das 1966). This work was then extended further and different aspects of such vesicle formation vis-à-vis their interpretation as a novel secretory mechanism of the cells were recorded and discussed in detail (Chatterjee and Das 1967). Those were indeed the first report

Fig. 1.2 *V. cholerae* cell
harvested from log phase
growth in peptone water,
transferred to glucose–saline
solution and incubated for
1.5 h. Electron micrograph
(*EM*) of ultrathin section
(cross-section) stained with
lead. Cell-wall bulging and
the formation of sac like
structures (*SS*) can be seen in
several places (*arrows*). Bar
represents 0.1μm [From
(Chatterjee and Das 1967)]

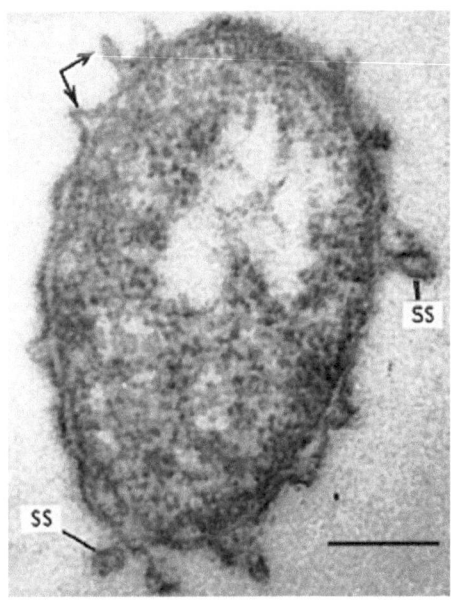

and systematic study of a Gram-negative bacterium releasing outer membrane
vesicles (OMVs) under normal growth conditions; further details are presented
hereafter.

1.3 Production of OMVs

The mode of formation of OMVs under normal growth conditions was first
demonstrated in *Vibrio cholerae* cells in the 1960s by electron microscopy
(Chatterjee and Das 1966, 1967). The cell wall, in places, was found to bulge out
and different stages of the bulging were recorded in the bacteria harvested from the
logarithmic growth phase (Fig. 1.3). When the wall region bulged out considerably, its neck became increasingly constricted and the membrane sac so formed
appeared to get pinched off the bacterial surface and released in the medium
(Fig. 1.4). Particles resembling the pinched off membrane sacs observed in thin
sections were also detected by electron microscopy in the bacteria-free filtrates of
the log phase cultures by metal shadowing (Fig. 1.5) and also by negative staining
techniques (Fig. 1.6). Identical particles were detected in the electron micrographs
by using any of the three preparation techniques of ultrathin sectioning, metal
shadowing, and negative staining. The trilamellar structure of the vesicle membrane, resembling that of the cell wall membrane, could be shown by the thin
sectioning technique (Fig. 1.7). The dimensions of these particles were found to
vary between 40 and 110 nm. No similar bulging process of the cell wall was

Fig. 1.3 *V. cholerae* cell harvested from log phase of growth in Syncase medium; uranyl acetate-stained. Chromatin strands (*CS*) are distinct in the nucleoplasm. Bulging of the cell wall is shown at A and B. Bar represents 0.1μm [From (Chatterjee and Das 1967)]

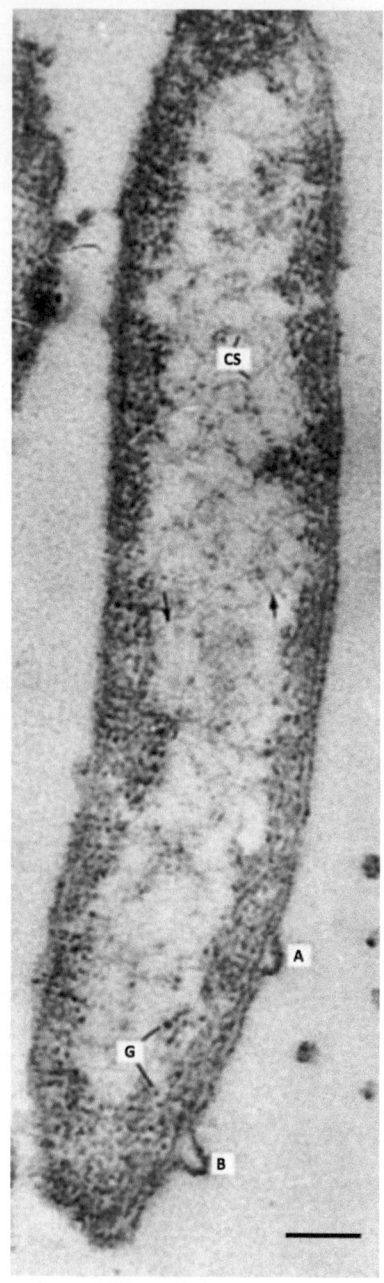

detected in vibrios harvested from the stationary growth phase or in vibrios undergoing plasmolysis (Fig. 1.8). It was further shown that at the initial plasmolysis stage, the vibrio protoplasm was found to retract from the cell wall but

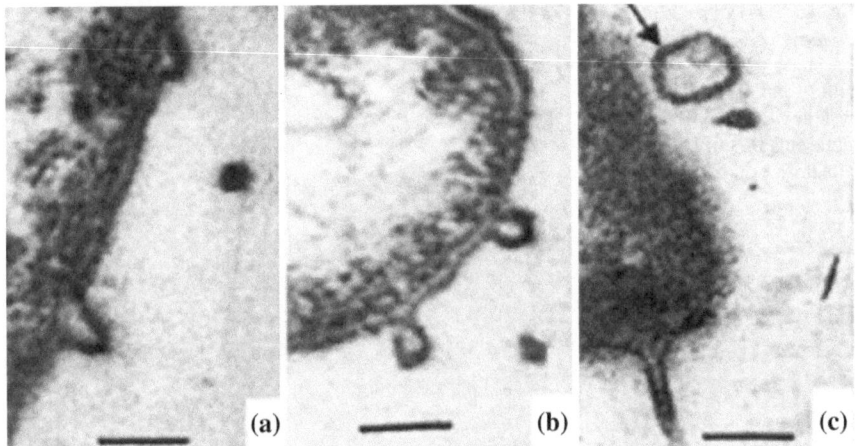

Fig. 1.4 Ultrastructural expressions (E.M. of ultrathin sections of actively growing cells) of the stages (a, b, and c) in the formation of extracellular membrane vesicles of *V.cholerae* cell. Bars represent 0.1 μm. [From (Chaudhuri and Chatterjee 2009)]

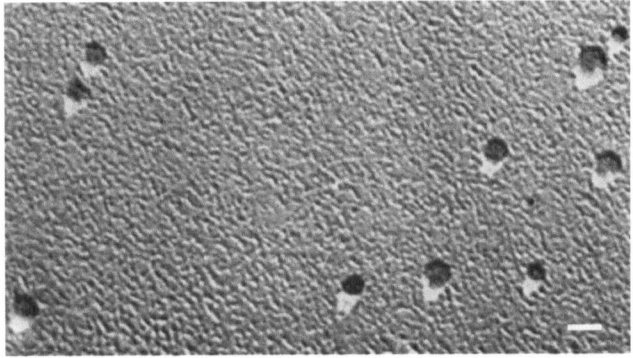

Fig. 1.5 OMVs derived from the Syncase culture filtrate of actively growing *V. cholerae* cells; electron microscopy of chromium-shadowed particles. Bar represents 0.1 μm [From (Chatterjee and Das 1967)]

the structure of the cell wall otherwise remained smooth. In all cases where cell wall excretion was noted in the vibrios, no evidence was obtained of any abnormal separation between the cell wall membrane and the plasma membrane or any bulging out of the protoplasm and the plasma membrane had remained undamaged. These observations seemed to rule out the possibility of any plasmolysis occurring during the preparation of the vibrios for electron microscopy. It was often observed that the bulged-out cell wall region was associated with some very small granules or particles at the surface (Fig. 1.9a). Similar granules (Fig. 1.9b) were also detected in the negatively stained preparations of the culture filtrates

Fig. 1.6 OMVs of *V.*
cholerae cells isolated from
log-phase peptone water
culture and negatively stained
by uranyl acetate. Bar
represents 0.1µm [From
(Chatterjee and Das 1967)]

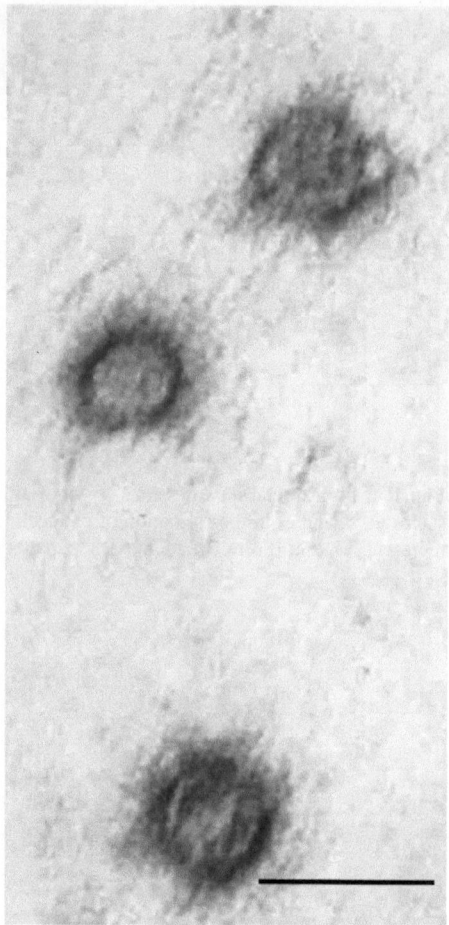

(Chatterjee et al. 1974). However, neither the chemical nor the functional identity
of these finer particles could be established by the authors.

1.4 A Novel Secretion Process

The authors (Chatterjee and Das 1967) observed that (1) the formation and release of
the surface blebs could not be seen in the resting phase vibrios or in vibrios
undergoing plasmolysis (Fig. 1.8), and (2) the membranes of some eukaryotic cells
were shown by different investigators to play a significant role in the secretion of
different materials including viruses. Accordingly, the authors (Chatterjee and Das
1967) interpreted and proposed that the release of blebs from the surface of actively
growing *V. cholerae* cells, as seen by electron microscopy, represented a novel

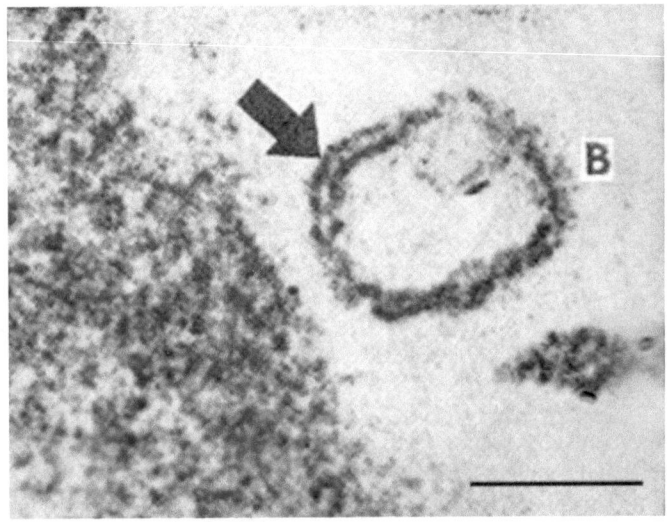

Fig. 1.7 E.M. of an OMV from *V. cholerae* cell in thin section. Trilamellar structure of the bounding wall is shown by *arrow*. Bar represents 0.1 μm [From (Chatterjee et al. 1974)]

Fig. 1.8 E.M. of *V. cholerae* cell (thin section) undergoing plasmolysis. The protoplast has retracted from the cell wall. Note that the cell wall structure is smooth and no bleb formation can be seen. Bar represents 0.1 μm [From (Chatterjee and Das 1967)]

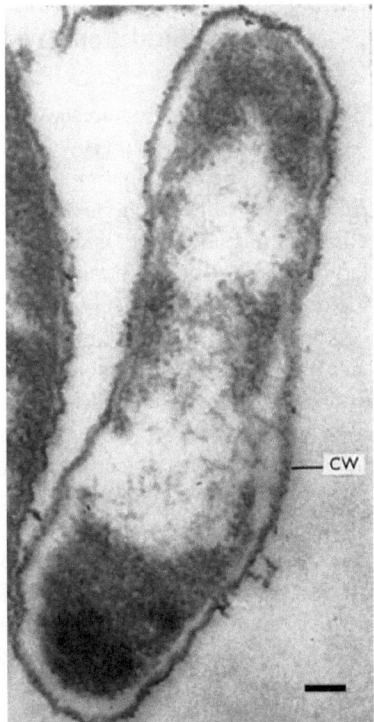

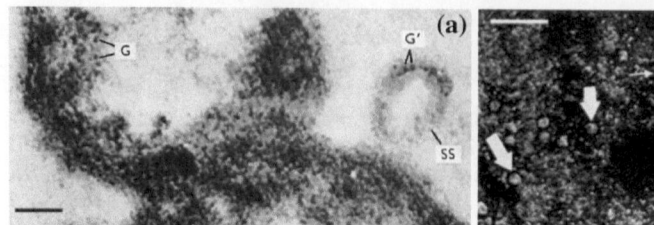
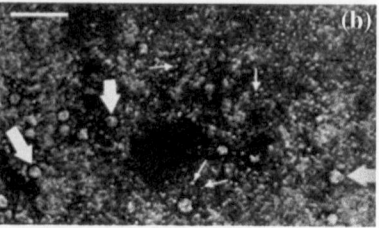

Fig. 1.9 a E.M. of a portion of *V. cholerae* (thin section) undergoing division. Presence of finer granules on the surface of the emerging OMV shown by arrow Bar represents 0.1μm [From (Chatterjee and Das 1967)] **b** Log-phase culture filtrate of *V. cholerae* Inaba 569B cells; electron micrograph of negatively stained preparation, showing bigger (OMVs, *thick arrows*) and much smaller particles (*thin arrows*) Bar represents 0.1μm [From (Chatterjee et al. 1974)]

secretory mechanism (to secrete non diffusible materials from the periplasm of the bacterial cell) of the young vibrios, a mechanism not reported till then to occur in any bacterial cell. This phenomenon is discussed further in Sect. 1.8 and Chap. 8 of this monograph along with the findings and observations of subsequent investigators.

1.5 Bleb Formation Under Abnormal Vis-à-Vis Normal Growth Conditions

Blebs or bleb like structures were shown by some investigators to originate from bacteria under abnormal growth conditions, a brief discussion of which seems to be relevant. While investigating the biochemistry of a lysine-requiring mutant strain of *Escherichia coli,* which was lacking the enzyme that decarboxylates diaminopimelic acid to lysine, Work and Denmann (1953) observed that such mutants excreted large amounts of diaminopimelic acid and to a lesser extent some other chemicals including lipids, polysaccharides, and proteins when grown under lysine-limiting conditions. Subsequently, a product termed lipoglycopeptide was isolated (Bishop and Work 1965) from the concentrated culture fluid after growth of this organism for 26 h in a well-aerated salts–glycerol medium containing 60-mg lysine HCl/liter. The authors, however, were at a loss to explain the observations. It was only when they took recourse to electron microscopy was the abnormal or unbalanced state of affairs of the bacteria in culture revealed (Knox et al. 1966; Work et al. 1966). After 4.5-h growth, the lysine-limited cells appeared normal (without any bleb on the surface). By 8 h when lysine was exhausted and growth had stopped, defects in the cell wall became apparent, small blebs being present. Blebs and a mass of stacked leaflets (their origin not explained by the authors) surrounding the bacteria became more prevalent after incubation had continued for 12 h or more when the bacteria were undergoing unbalanced growth and/or decay. Other changes taking place during this time included a relatively wide electron-transparent zone between the multilayered cell wall and the cytoplasmic membrane and circular areas of low electron density in the cytoplasm, acknowledging thereby that the bacteria were in a state of abnormal growth or decay. It may be noted that a similar wide electron transparent zone between the cell wall and plasma membrane was noted in *V. cholerae* cells subjected to plasmolysis (Fig. 1.8; Chatterjee and Das 1967). It was, however, suggested that the extracellular lipoglycopeptide was produced by the formation and release of blebs or bleb like structures from the cell surface (Knox et al. 1966, Work et al. 1966). The changes observed by these authors in lysine-requiring *E. coli* growing in lysine-deficient medium were thus related to an abnormal or decaying state of the bacteria and could not, in any case, be explained as a well-regulated physiological process of the cells under S.O.S. In comparison, the observations of Chatterjee and Das represented an inherent property of normally growing *V. cholerae* cells, a property till then unreported to occur in any bacterial cell (Chatterjee and Das 1966, 1967).

1.6 A Temporary Setback

Kennedy and Richardson subsequently reported that they did not find any bulging out and pinching off of portions of the cell wall membrane of *V. cholerae* during their active phase of growth (Kennedy and Richardson 1969). These authors further showed that with proper fixation, the ribosome content of the cells seemed to be reduced, some well-preserved spheroplasts were present, and the nuclear area increased especially during the late logarithmic phase. They interpreted that the observations of Chatterjee and Das (1966, 1967) probably resulted from poor fixation or improper choice of fixative used for electron microscopy. In support of their observations they showed that during a suboptimal fixation, vesicle production was observed throughout the growth cycle, which reached their greatest number in stationary phase of growth.

   For some subsequent years the artifactual origin of the blebs appearing on the surface of bacteria, *V. cholerae* in particular (Chatterjee and Das 1966, 1967), was seriously believed. Furthermore, in heat-treated *E. coli* cells, bleb or bleb like structures were shown to form at places by detachment of the cell wall membrane from the underlying peptidoglycan layer and by "ballooning" of the cell wall membrane (De Petris 1967). This report along with the two aforesaid publications (Knox et al. 1966, Work et al. 1966), on some lysine-requiring mutants of *E. coli* producing blebs or bleb like structures when they were leaking and undergoing unbalanced growth and decay, lent further credence to the ideas of Kennedy and Richardson (1969).

   In this context, a brief discussion of some other contemporary and relevant publications is thought desirable in order to dispel any idea of their relation to bleb formation by bacteria under normal growth. While studying the surface structures of *E. coli,* Bayer and Anderson used unfixed cells which were quick-frozen at −190 °C, cut open at −30 °C, and then negatively stained for electron microscopy after the cytoplasm had been allowed to escape (Bayer and Anderson 1965).

The authors revealed some structures at the cell surface, which were treated with different chemicals to learn their chemical identity. Neither the authors claimed, nor did the electron micrographs reveal, the formation of any OMV from the surface of cells devoid of any cytoplasm. In another study, Bladen and Waters resolved the multilayered structure of the cell wall and the plasma membrane of some strains of *Bacteroids*, the inward annular growth of the cytoplasmic membrane and did not reveal or claim to have shown the formation of OMVs either during normal growth or after plasmolysis of the cells (Bladen and Waters 1963). On the other hand, Bladen and Mergenhagen examined the ultrastructure of normal, phenol-water extracted, and lysozyme-treated *Veillonella* cells and revealed the endotoxic activity of the outer three-related to the formation of blebs layered membrane (Bladen and Mergenhagen 1964). Again, the authors did not claim to have shown the formation of OMVs on the surface of treated or untreated cells. Any reference to these publications as by actively growing bacteria is likely to confuse our under-standing of bacterial OMVs. In fact, following the properties of the lipid molecules, blebs or bleb like structures will automatically form when some portion of the bacterial outer membrane is snatched away by chemical or mechanical action.

In the context of all these reports, the entire process of bleb formation was then re-examined or reinvestigated by Chatterjee and co-workers (Chatterjee et al. 1974; Chatterjee and Sur 1974) as discussed below.

1.7 Further Work on Bleb Formation

A critical re-examination of the entire matter of surface bleb formation in *V. cholerae* cells using different culture media (Syncase and peptone water media), fixatives, and embedding media for electron microscopy and also biochemical monitoring of the growth and macromolecular synthesis ascertained that bleb formation did take place in the logarithmic phase irrespective of culture medium, fixative, or embedding medium used and that it was not accompanied by any cell lysis or any abnormal separation between the cell wall and plasma membrane of the cell (Chatterjee et al. 1974; Chatterjee and Sur 1974). An interesting and constant feature of the actively growing cells was the presence of finer particles lying in association with the cell wall or surface blebs or lying immediately outside the cell surface where apparently no bleb like structure could be seen.

Growth curves of different strains of *V. cholerae* in different culture media used were rather typical of normal bacterial growth (Chatterjee et al. 1974). Unlike the observations of Kennedy and Richardson (1969), the optical density of the culture did not fall significantly during the early hours of the stationary phase, the spheroplasts could not be detected at any stage of logarithmic growth, and the macromolecular synthesis continued to increase until the logarithmic growth ended. Furthermore, no decrease of the cell's ribosomal content during the logarithmic growth phase could be seen. Also no bleb formation took place, unlike the observations of Kennedy and Richardson (1969), during the stationary or resting phase of growth (Chatterjee et al. 1974).

1.8 Confirmation of the Discovery by Contemporary Investigators

The mechanism of bleb formation reported to occur in *V. cholerae* as a novel secretory process (Chatterjee and Das 1966, 1967) was quickly confirmed and reported to take place under normal growth conditions in different bacteria by many investigators (Avakian et al. 1972; Devoe and Gilchrist 1973; Farkas-Himsley et al. 1971; Kushnarov 1971; Pike and Chandler 1975). Avakyan et al. (1972), were in fact, the first to report the production of OMVs by *V. cholerae* growing inside a host. Pike and Chandler (1975) found the release of somatic antigens into the culture fluid of *V. cholerae* and interpreted that this could happen by the bulging and pinching off of the cell-wall materials as observed by Chatterjee and Das (1967) and was not necessarily a consequence of cell lysis. In addition, it would be interesting to reproduce a few lines from the paper of Farkas-Himsley et al. (1971) as: In an electron microscopic study of *V. cholerae* (Chatterjee and Das 1967) similar bulges, presumably originating from cell wall material, were described...These authors suggested that the process of pinching off of bulges was intimately related to an active secretory mechanism in vibrios, simulating reverse pinocytosis (Palade 1961), whereby non-diffusible substances could be removed from the cell (Chatterjee and Das 1967). Indeed, in line with the above, our non-induced control vibrio cells were found to undergo this transition to a smooth outer membrane.... Similar observations on the release of blebs from a bacterial surface under normal conditions were subsequently recorded by many other investigators in different Gram-negative bacteria (Kuehn and Kesty 2005). The released blebs or OMVs have now been found to play distinctive roles in many different areas of biology and medicine, a detailed story of which forms, in fact, the subject matter of this monograph.

1.9 Essential Findings of the Discovery Vis-à-Vis Later Works

In a nutshell, (1) the details of the mode of formation and release of blebs or vesicles or OMVs by a Gram-negative bacterium under normal growth condition were recorded for the first time and (2) it was proposed, by comparison with the observed role of membranes in the secretory processes of some eukaryotic cells, that the release of these blebs in the extracellular medium represented a novel secretory mechanism of the actively growing bacterial cells, and was not due to the lysis of the cells (Chatterjee and Das 1966, 1967). These two findings were subsequently confirmed and established by different investigators working with many different Gram-negative bacteria. McBroom and Kuehn (2005) and Kuehn and Kesty (2005) subsequently established that the production of OMVs represented a novel secretion mechanism of bacteria, thereby supporting the original proposal of Chatterjee and Das (1966,1967) and providing additional supporting data and critical analysis (Kuehn and Kesty 2005).

Although the presence of very small particles (which dimensionally may represent the cholera toxins, the molecular weight of which was by then fairly well known), was shown to be associated with the blebs, the particles' nature and function could not be identified (Chatterjee and Das 1967). Whether these particles represented the choleragenic toxins was at that time only a matter of conjecture. However, it is of interest to note in this respect that the extracellularly released toxin particles were shown, many years later, to be associated with the blebs (now known as OMVs) released by enterotoxigenic *E. coli* (Horstman and Kuehn 2002) and more recently by *V. cholerae* (Chatterjee and Chaudhuri 2011) cells.

Furthermore, it was established years later that vesiculation is not a consequence of bacterial lysis or disintegration of the bacterial envelope and cannot be correlated with membrane instability and that vesiculation is an important process in the growth of Gram-negative bacteria (McBroom et al. 2006; Mug-Opstelten and Witholt 1978; Yaganza et al. 2004; Zhou et al. 1998). Therefore, it is important to differentiate such unnatural bleb formation under abnormal or unbalanced growth (as in lysine- or nutrient-deficient media) or growth under antibiotic (such as gentamicin) treatment from those produced under natural conditions (natural blebs) as an inherent property of the bacteria. The blebs or bleb like particles produced under such abnormal conditions (termed unnatural blebs or OMVs) were shown to contain, in addition, cytosolic components in their lumen (as evidence of leaky membranes or cell lysis) in addition to the parent bacteria having an abnormal (indicative of plasmolysis) structure (Knox et al. 1966; Work et al. 1966). The unnatural OMVs are discussed in this monograph as and when required for relevance only.

References

Avakian AA, Sinel'nikova MP, Pereverzev NA, Gurskii IuN (1972) Electron microscopic study of biopsied sections of small intestine mucosa in patients with cholera and characteristics of the ultrastructure of causative agents of cholera in relation to toxinogenesis. Zh Mikrobiol Epidemiol Immunobiol 49:86–92

Bayer ME, Anderson TF (1965) The surface structure of Escherichia coli. Proc Natl Acad Sci U S A 54:1592–1599

Bishop DG, Work E (1965) An extracellular glycolipid produced by Escherichia coli grown under lysine-limiting conditions. Biochem J 96:567–576

Bladen HA, Mergenhagen SE (1964) Ultrastructure of veillonella and morphological correlation of an outer membrane with particles associated with endotoxic activity. J Bacteriol 88: 1482–1492

Bladen HA, Waters JF (1963) Electron microscopic study of some strains of bacteroides. J Bacteriol 86:1339–1344

Chatterjee D, Chaudhuri K (2011) Association of cholera toxin with Vibrio cholerae outer membrane vesicles which are internalized by human intestinal epithelial cells. FEBS Lett 585:1357–1362

Chatterjee SN, Das J (1966) Secretory activity of Vibrio cholerae as evidenced by electron microscopy. In: Uyeda (ed) Electron Microscopy 1966. Maruzen Co. Ltd, Tokyo

Chatterjee SN, Das J (1967) Electron microscopic observations on the excretion of cell-wall material by Vibrio cholerae. J Gen Microbiol 49:1–11

Chatterjee SN, Sur P (1974) Surface blebs on Vibrio cholerae cells. In: Saders JV, Goodchild D (eds) Canberra, Australia, pp 652–653

Chatterjee SN, Adhikari PC, Maiti M, Chaudhuri CR, Sur P (1974) Growth of Vibrio cholerae cells: biochemical and electron microscopic study. Indian J Exp Biol 12:35–45

Chaudhuri K, Chatterjee SN (2009) Cholera toxins. Springer, Heidelberg

De SN (1959) Enterotoxicity of bacteria-free culture-filtrate of Vibrio cholerae. Nature 183:1533–1534

De Petris S (1967) Ultrastructure of the cell wall of Escherichia coli and chemical nature of its constituent layers. J Ultrastruct Res 19:45–83

Devoe IW, Gilchrist JE (1973) Release of endotoxin in the form of cell wall blebs during in vitro growth of Neisseria meningitidis. J Exp Med 138:1156–1167

Farkas-Himsley H, Kormendy A, Jayawardene A (1971) Electron microscopy of Vibrio comma during vibriocin production. Cytobios 3:97–116

Horstman AL, Kuehn MJ (2002) Bacterial surface association of heat-labile enterotoxin through lipopolysaccharide after secretion via the general secretory pathway. J Biol Chem 277: 32538–32545

Kennedy JR, Richardson SH (1969) Fine structure of Vibrio cholerae during toxin production. J Bacteriol 100:1393–1401

Knox KW, Vesk M, Work E (1966) Relation between excreted lipopolysaccharide complexes and surface structures of a lysine-limited culture of Escherichia coli. J Bacteriol 92: 1206–1217

Kuehn MJ, Kesty NC (2005) Bacterial outer membrane vesicles and the host-pathogen interaction. Genes Dev 19:2645–2655

Kushnarov VM (1971) Mikrobiologogie 40:918–923

McBroom AJ, Johnson AP, Vemulapalli S, Kuehn MJ (2006) Outer membrane vesicle production by Escherichia coli is independent of membrane instability. J Bacteriol 188:5385–5392

Mug-Opstelten D, Witholt B (1978) Preferential release of new outer membrane fragments by exponentially growing Escherichia coli. Biochim Biophys Acta 508:287–295

Palade GE (1961) The secretory process of the pancreatic cell. In: Boyd JD, Johnson FR, Lever JD (eds) Électron microscopy in anatomy. Arnold, London

Pike RM, Chandler CH (1975) Partial purification and properties of somatic antigen spontaneously released from Vibrio cholerae. Infect Immun 12:187–192

Work E, Denman RF (1953) The use of a bacterial culture fluid as a source of alpha-diaminopimelic acid. Biochim Biophys Acta 10:183

Work E, Knox KW, Vesk M (1966) The chemistry and electron microscopy of an extracellular lipopolysaccharide from Escherichia coli. Ann N Y Acad Sci 133:438–449

Yaganza ES, Rioux D, Simard M, Arul J, Tweddell RJ (2004) Ultrastructural alterations of Erwinia carotovora subsp. Atroseptica caused by treatment with aluminum chloride and sodium metabisulfite. Appl Environ Microbiol 70:6800–6808

Zhou L, Srisatjaluk R, Justus DE, Doyle RJ (1998) On the origin of membrane vesicles in gram-negative bacteria. FEMS Microbiol Lett 163:223–228

Chapter 2
Gram-Negative Bacteria: The cell Membranes

Abstract This chapter presents a brief outline of our current knowledge of the structures of the bounding membranes (the inner and the outer membranes and the intermediate periplasmic layer) of a Gram-negative bacterial cell. Also the structure and chemical composition of the outer membrane vesicles (OMVs) originating from the surface of these bacteria including their proteomic profile, as obtained mainly by mass spectroscopic and related studies, have been presented in brief.

Keywords Inner membrane · Outer membrane · Peptidoglycan · OMVs · Structure · Chemical composition · Mass spectrometry · Protein profile

2.1 Inner and Outer Membranes

The Gram-negative bacteria are usually bounded by two membranous structures (Fig. 2.1). The inner one (IM), called the plasma membrane, is a trilamellar structure that bounds the bacterial protoplasm and is composed of a phospholipids bilayer. The outer membrane (OM) also presents a trilamellar structure (with two electron dense leaflets, outer and inner) in the electron micrograph and consists of proteins, including porins, receptors, and an asymmetric distribution of lipids. The outer leaflet is composed primarily of lipopolysaccharide (LPS) projecting outside and the inner leaflet containing phospholipids and lipoproteins. The LPS of a Gram-negative bacterium consists of three different sectors: (i) lipid-A, (ii) the core polysaccharide comprising the inner and the outer cores, and (iii) the O-specific polysaccharide chains (Fig. 2.1) projecting outward. The lipid portion of LPS serves as the lipid anchor and is commonly composed of fatty acids, sugars, and phosphate groups. The chemical structures of lipid-A, core polysaccharide, and O-specific polysaccharide chains of *Vibrio cholerae* are shown in Fig. 2.2 a, b,

S. N. Chatterjee and K. Chaudhuri, *Outer Membrane Vesicles of Bacteria*,
SpringerBriefs in Microbiology, DOI: 10.1007/978-3-642-30526-9_2,
© The Author(s) 2012

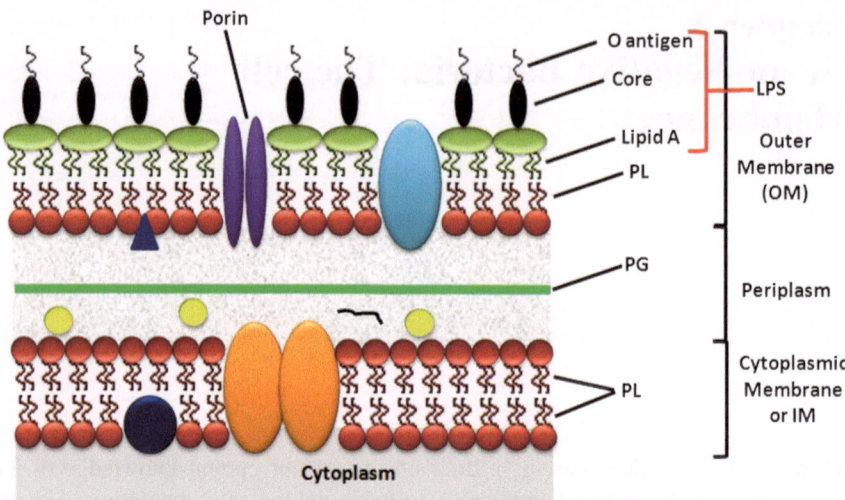

Fig. 2.1 Schematic diagram of the outer membrane (*OM*), cytoplasmic or *IM,* and the intermediate periplasmic layer containing the peptidoglycan (*PG*). The *IM* consists of the two phospholipid (*PL*) leaflets and different lipoproteins. The outer membrane consists of two leaflets, the inner leaflet being composed of one phospholipid layer and the outer leaflet of lipid-A, core polysaccharide, and the O-antigen polysaccharide chains projecting outward

and c. These two membranes, IM and OM, are separated by a gel-like layer known as the periplasm. The periplasm contains a thin layer ($\sim$4- nm thick) of peptidoglycan (PG) which is connected to the OM and also the inner membrane (IM) through different protein–protein interactions and other proteins including the so-called resident "housekeeping" proteins and enzymes, the resident and transient components of secretory pathways and the like. In *Pseudomonas aeruginosa* there are three lipoproteins, Oprl, OprL, and OprF, that connect the OM with the PG layer. In wild- type *Salmonella spp* there are specific domains in the envelope that promote interactions between the outer membrane protein (Omp) and the peptidoglycan layer (PG) and also interactions between the Omp and the inner membrane protein (IMP) involving the PG; (Deatherage et al. 2009). Such protein–protein interactions involving the PG are primarily responsible for giving the required strength and stability to the bacterial envelope or rather the surface structures of bacteria.

A schematic description of the presence and interactions between these different proteins in wild- type *Salmonella spp* is presented in Fig. 2.3. The integral proteins in the OM include OmpC, OmpF, OmpX, and NmpC; those involved in the interactions between the OM and the PG include OmpA, LppA, and LppB, and the peptidoglycan- associated lipoprotein Pal in the OM can interact with TolA in the IM either directly or via the periplasmic protein, TolB. Pal can also interact directly with the PG. The OM contains the unique trimeric proteins known as porins. Porins are channel-forming proteins that allow small

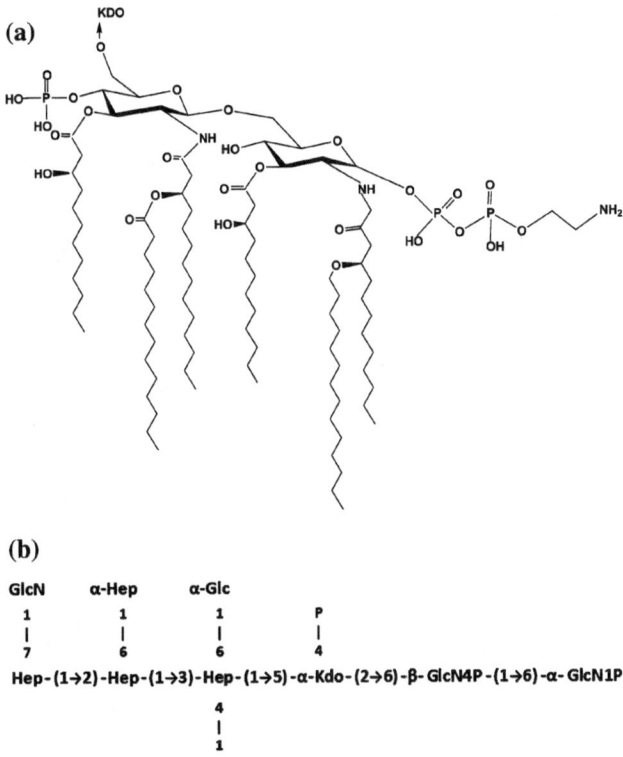

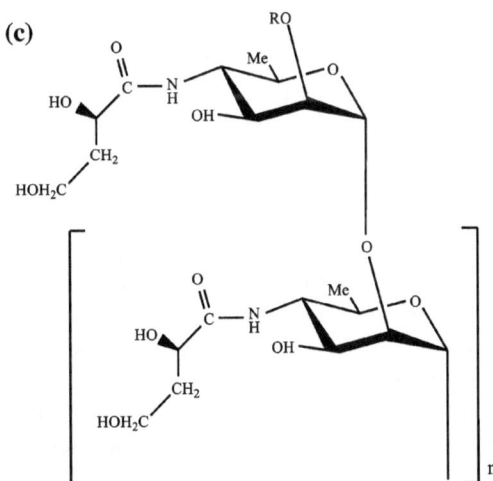

Fig. 2.2 **a** Chemical structure of lipid-A of *V. cholerae* O1. **b** Chemical structure of the core polysaccharide (core-PS) of *V. cholerae* O1 strain 95R. **c** Chemical structure of the O-PS of *V. cholerae* O1. The O-PS structures of the two serotypes, Inaba and Ogawa, are the same except at the position O-2 of the upstream, terminal perosamine group; R = CH$_3$ in Ogawa strain and R = H only in Inaba strain; *n*, represents the number of repeating units, which may be between 12 and 18 Chatterjee and Chaudhuri (2003); and Chaudhuri and Chatterjee (2009)

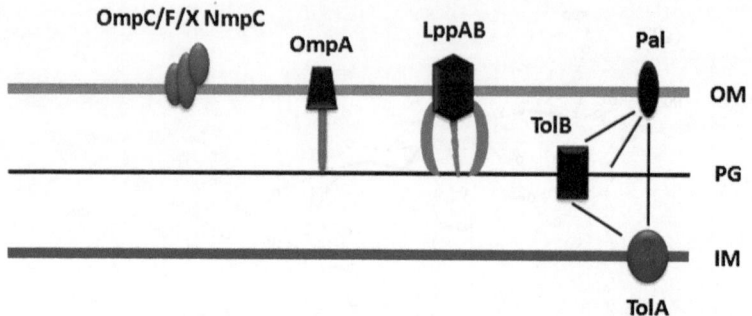

Fig. 2.3 Proteins interconnecting different layers of the cell envelope. Integral OM proteins OmpC, OmpF, OmpX, and NmpC normally do not connect to other layers of the envelope. Lpp and OmpA form an interconnection between PG and OM, whereas Pal of the OM, Tol B of the PG, and Tol A of the IM form interconnections between the different components of the envelope. The thinner straight line represents covalent interaction; the thicker straight line or the curved lines represent noncovalent interactions

molecules (<600 daltons) to pass through and enter the periplasmic space. Once in the periplasm, proteins within the plasma membrane allow transport of molecules into the cytoplasm. In *P. aeruginosa*, a specific porin, OprF, not only allows passage of small molecules, but is also associated with the underlying PG within the periplasm. Lipoproteins are also present in the periplasm. The structure of the OM of Gram-negative bacteria plays a dynamic role in the formation of outer membrane vesicles (OMVs).

2.2 Structure of OMVs

The OMVs originate by a process of bulging out and pinching off of a portion of the bacterial OM (Chatterjee and Das 1966, 1967) thereby entrapping much of the materials of the underlying periplasm. However, inclusion of periplasmic materials into the OMVs is dependent on some specific sorting mechanism. They are spherical in shape (Chatterjee and Chaudhuri 2011) with sizes varying between 50 and 250 nm as reported by most researchers (Fig. 2.4;) (Beveridge 1999; Mashburn-Warren and Whiteley 2006). In our experience, OMVs of sizes significantly smaller than 50 nm have been found and these are often lost during the procedures for isolation of OMVs. The OMVs are bounded by a trilamellar structure similar to that of the bacterial OM (Figs. 1.7 and 2.5). The bounding membrane of the OMVs also has a similar chemical structure to that of the bacterial OM and accordingly contains the antigenic LPS projecting outside.

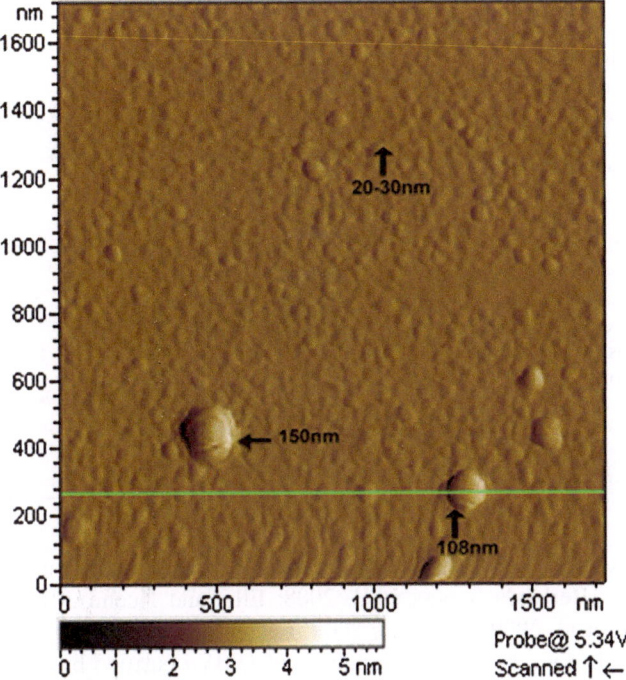

Fig. 2.4 Atomic force microscopy image of air-dried *V. cholerae* OMVs in amplitude mode (Bar = 200 nm). [From (Chatterjee and Chaudhuri 2011)]

2.3 Chemical Composition of OMVs

The OMVs originate from the bacterial surface and are, generally, known to contain OMPs, many of the periplasmic proteins, phospholipids, and LPS and other factors associated with virulence (Horstman and Kuehn 2000; Kuehn and Kesty 2005; Wai et al. 2003). By using the two-dimensional SDS-PAGE and MALDI-TOF mass spectrometric techniques, the major OMV proteins identified in *Salmonella sp.* were OmpC, OmpF, NmpC, OmpX, OmpA, LppA, LppB, Pal, and Tol B (Deatherage et al. 2009). During their formation and release from the bacterial surface, the OMVs entrap some of the underlying periplasmic constituents which may vary with bacterial growth conditions and for different bacteria. The OMVs from different bacteria can entrap toxins, enzymes, DNA, adhesins, and other virulence factors (Ellis and Kuehn 2010). Table 2.1 presents the names of different virulence factors carried by the OMVs in their lumens and the corresponding bacterial species from which they originated. The overall chemical composition of OMVs thus depends on various factors controlling growth and the species from which they originate. Recent evaluations showed that almost all OMV preparations were enriched in envelope components (Kuehn and Kesty 2005). Some of the preparations, however, were also found to contain a small

Fig. 2.5 Schematic diagram of an OMV showing the different possible luminal components (proteins, double- stranded DNA, RNA, plasmid, etc.), and the trilamellar structure of the vesicle membrane including the outwardly projecting O-PS chains, the proteins spanning the two leaflets of the membrane, and some toxin particles either bound to the LPS chains or near the outer surface of the membrane or within the lumen

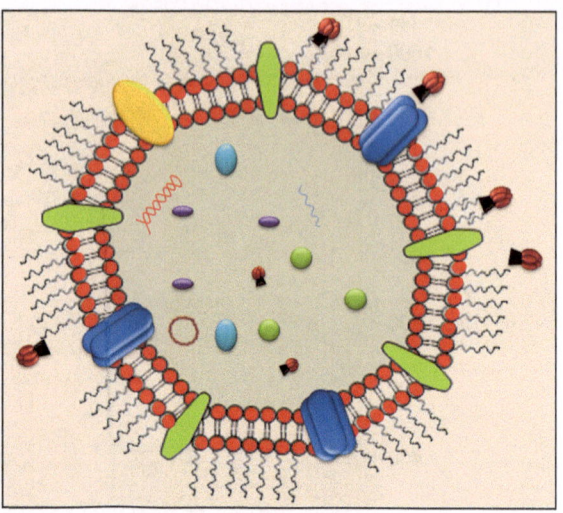

amount of cytosolic and IM proteins, the basis of which remains unclear or rather controversial (Berlanda Scorza et al. 2008; Ellis and Kuehn 2010; Galka et al. 2008; Kwon et al. 2009; Lee et al. 2007; Lee et al. 2008; Sidhu et al. 2008; Xia et al. 2008). In fact, biochemical analysis of OMVs purified by density gradient centrifugation revealed that they consisted only of the proteins and lipids of the OM and periplasm and did not contain any IM and cytoplasmic components (McBroom and Kuehn 2005). Pathogenic bacteria were shown to release OMVs containing adhesins, toxins, and immunomodulatory compounds. Because of their lipid contents, such vesicles were found to fractionate into lighter density fractions than solubly secreted proteins (Allan and Beveridge 2003; Allan et al. 2003; Dorward and Garon 1989; Dorward et al. 1989; Horstman and Kuehn 2000).

2.4 Proteomic Profile of OMVs

2.4.1 Isolation and Purification

In order to have a dependable proteomic profile of OMVs of any Gram-negative bacteria, the OMVs are first required to be isolated and highly purified so that no trace of materials released from bacterial lysis or any nonvesicular component contaminates the preparation. For this, no single method of isolation and purification can serve the purpose and a combination of differential centrifugation to remove cell debris and whole cells and ultracentrifugation to pellet the OMVs is the minimum requirement (Wai et al. 2003). Furthermore, filtration of the cell culture supernatant through membrane filters (0.22–0.45 μm) before ultracentrifugation may ensure better elimination of any contamination (Berlanda Scorza et al. 2008;

Table 2.1 Virulence factors associated with OMVs derived from various bacterial species

Bacterial spp.	Associated disease	Virulence factors	Activity	Reference
Actinobacillus pleuropneumoniae	Porcine contagious pleuropneumonia	Apx toxin	Hemolytic activity, cytotoxicity	Negrete-Abascal et al. (2000)
		Proteases	Proteolytic activity, host damage	
Actinobacillus actinomycetemcomitans	Periodontal disease	Leukotoxin	Pore-forming and membranolytic activity	Goulhen et al. (1998), Karched et al. (2008), Kato et al. (2002)
		Lipopolysaccharide (LPS)	Endotoxic activity	
		GroEL	Cytotoxicity	
		Peptidoglycan associated lipoprotein (PAL)	Proinflammatory activity on human whole blood	
Bacteroides fragilis	Colon inflammation/ Colon tumor	Hemagglutinin	Hemagglutination	Patrick et al. (1996)
		Alkaline phosphatase	Enzymatic activities causing host damage	
		Esterase lipase		
		Acid phosphatase		
		Phosphohydrolase		
		α- and β- galactosidases		
		α-glucosidase		
		Glucosaminidase		
		β-glucuronidase		
Bacteroides succinogenes	Abscesses, bacteremias	Cellulase	Aryl-β-glucosidase, endoglucanase	Forsberg et al. (1981)
		Xylanase	Aryl-β-xylosidase, xylanase activities	

(continued)

Table 2.1 (continued)

Bacterial spp.	Associated disease	Virulence factors	Activity	Reference
Bordetella pertussis	Whooping cough	Adenylate cyclase hemolysin	Cytotoxicity	Hozbor et al. (1999)
		Filamentous hemaglutinin (FHA)	Adhesion, agglutinates erythrocytes	
		Pertussis toxin (Ptx)	Inhibition of eukaryotic adenylate cyclase, increases cAMP	
Borrelia burgdorferi	Lyme disease	Outer surface proteins A and B (OspA, OspB) OspD	Adherence to host cells and tissue	Dorward et al. (1991), Shoberg and Thomas (1993)
		Decorin-binding protein A (DbpA)		
Brucella melitensis	Brucellosis	Outer membrane proteins Omp25, Omp31	ND	Gamazo and Moriyon (1987)
Burkholderia cepacia	Respiratory tract infection especially in cystic fibrosis patients	Nonhemolytic phospholipase C (PLC-N)	Lipolytic activity	Allan et al. (2003)
		Lipase		
		Pseudomonas cepacia protease (PSCP)	Protease activity	
		40-kDa protease		
Campylobacter jejuni	Gastroenteritis	Cytolethal discending toxin (CDT)	Genotoxicity	Lindmark et al. (2009)
Enterohemorragic *E. coli* (EHEC)	Bloody diarrhea, hemolytic-colitis	Cytolysin (ClyA)	Pore formation, membranolytic activity	Kolling and Matthews (1999), Wai et al. (2003)
		Shiga toxin	Cytotoxic, inhibit protein synthesis	

(continued)

Table 2.1 (continued)

Bacterial spp.	Associated disease	Virulence factors	Activity	Reference
Enterotoxigenic *E. coli* (ETEC)	Diarrhea	Heat labile enterotoxin (LT)	Increases adenylate cyclase and cAMP, loss of fluid and electrolyte in host	Horstman and Kuehn (2000)
Extraintestinal pathogenic *E. coli* (ExPEC)	Extraintestinal infections such as urinary tract infections, neonatal meningitis, septicemia	Alpha-hemolysin	Hemolytic, causing detachment of cells from monolayer	Balsalobre et al. (2006)
		Cytolethal descending toxin (CDT)	Genotoxicity	
		Iron and hemin binding OMPs	Iron acquisition	
Shiga toxin producing *E. coli* (STEC)	Hemorrhagic colitis,	Shiga toxin	Inhibition of protein synthesis and death of host cells	Kolling and Matthews (1999), Yokoyama et al. (2000)
Uropathogenic *E. coli* (UPEC)	Urinary tract infection	Cytotoxic necrotizing factor type 1 (CNF1)	Cytotoxicity	Kouokam et al. (2006)
Helicobacter pylori	Gastritis and peptic ulcer, promotes gastric cancer	Vacuolating cytotoxin (VacA)	Vacuolating activity	Hynes et al. (2005), Keenan et al. (2000)
		Urease	Hydrolyzes urea	
		Helicobacter cysteine-rich proteins (Hcp)	Interferes with host cell functions	
		Lewis antigen LPS	Cytotoxic, stimulating proliferation, IL-8 secretion	

(continued)

Table 2.1 (continued)

Bacterial spp.	Associated disease	Virulence factors	Activity	Reference
Legionella pneumophila	Legionnaires' disease	Mip (lpg0791), Macrophage infectivity potentiator	Inhibition of autophagy in macrophages	Fernandez-Moreira et al. (2006), Galka et al. (2008)
		Flagellin	Evasion and spreading in host	
		LaiE/LaiF	Adhesion to and invasion of human lung epithelial cell	
		Intracellular multiplication protein K and X (IcmK/IcmX)	Involved in secretion and intracellular replication of Legionella in macrophages	
		Phospholipase C		
		Acid phosphatases		
		Diphosphohydrolase		
		Chitinase	Glycosylase, promotes persistence in the lung	
		Proteases	Interfere with immune function	
		Hsp60	Involved in adherence and invasion	
Moraxella catarrhalis	Otitis media and sinusitis, occasional cause of laryngitis	Ubiquitous surface protein A1 and A2 (UspA1/UspA2)	Binds C3 complement in serum	Tan et al. (2007)
Myxococcus xanthus	Nonpathogenic	TonB transporters	Predatory behavior and multicellularity	Kahmt et al. (2010)

(continued)

Table 2.1 (continued)

Bacterial spp.	Associated disease	Virulence factors	Activity	Reference
Neisseria gonorrheae	Sexually transmitted disease gonorrhea	Porin B	Serum resistance, B-cell activation	Zhu et al. (2005)
Neisseria meningitidis	Meningococcal disease	PorA, PorB	Adherence to host cells, serum complement resistance	Bjerre et al. (2000), Masignani et al. (2003), Schlichting et al. (1993), Vipond et al. (2006)
		NlpB	Lipoprotein, maintains membrane integrity	
		NarE	Potent toxin, ADP-ribosyltransferase activity	
Photorhabdus luminescens	Insect pathogen	Toxin AB GroEL	Insecticidal activity Cytotoxicity	Guo et al. (1999), Khandelwal and Banerjee-Bhatnagar (2003)
Porphyromonas gingivalis	Periodontal disease, gingivitis	Arg- and Lys-gingipain cysteine proteinases	Hemoglobinase activity	Duncan et al. (2004), Grenier (1992), Grenier and Mayrand (1987), Kamaguchi et al. (2003)
		Fimbriae	Interfere with immune response	

(continued)

Table 2.1 (continued)

Bacterial spp.	Associated disease	Virulence factors	Activity	Reference
P. aeruginosa	Lung infection of patients with chronic obstructive pulmonary disease (COPD), pneumonia, cystic fibrosis (CF), and bronchiectasis	Phospholipase C	Decrease of apical CFTR expression	Bauman and Kuehn (2006), Ciofu et al. (2000), Cota-Gomez et al. (1997), Kadurugamuwa and Beveridge (1995), Li et al. (1998), MacEachran et al. (2007), Mashburn and Whiteley (2005), Bomberger et al. (2009)
		Hemolysin	Lyse red blood cells	
		Alkaline phosphatase	In vitro enzyme activities	
		Cystic fibrosis transmembrane regulator (CFTR) inhibitory factor (Cif)	Degradation of CFTR	
		2-heptyl-3-hydroxy-4-quinolone (PQS)	bactericidal	
		Murein hydrolases	Lysis of peptidoglycan	
		Protease	Proteolytic activity	
		β-lactamase	IL-8 stimulation	
Salmonella typhi	Enteric fever typhoid	Cytolysin (ClyA)	Pore formation, membranolytic activity	Wai et al. (2003)
Salmonella typhimurium	Diarrheal disease	OmpC	ND	Bergman et al. (2005), Yoon et al. (2011)
		Protective antigens	ND	
Shigella dysenteriae serotype 1	Shigellosis (bacillary dysentery)	Shiga toxin 1	Toxicity	Dutta et al. (2004)
Shigella flexneri	Diarrhea	IpaB, IpaC, IpaD	Invasion	Kadurugamuwa and Beveridge (1998)
Treponema denticola	Periodontal diseases	Dentilysin	Chymotryptic activity	Chi et al. (2003), Rosen et al. (1995)
		Adhesins	Adherence to host	
		Proteases	Disruption of tight junctions	

(continued)

Table 2.1 (continued)

Bacterial spp.	Associated disease	Virulence factors	Activity	Reference
Vibrio anguillarum	Fish pathogen	Metalloprotease	Protease, metalloprotease	Hong et al. (2009)
		Hemolysin	Hemolytic activities	
		Phospholipase	Lipolytic activity	
Vibrio cholerae	Diarrhea	RTX toxin	Cell rounding, depolymerizing actin	Boardman et al. (2007), Chatterjee and Chaudhuri (2011)
		Cholera Toxin (CT)	cAMP activation, fluid accumulation	
Xanthomonas campestris	Plant pathogen	Cellulase	ND	Sidhu et al. (2008)
		β-glucosidase		
		Type 3 secretion system proteins		
		xylosidase		
		avirulence proteins		
Xenorhabdus nematophilus	Insect pathogen	Bacteriocin	Insecticidal	Khandelwal and Banerjee-Bhatnagar (2003)
		Fimbrial adhesin	Adherence to host	
		Pore-forming toxin	Cytotoxicity	
		Chitinase	Chitinase activity	

Ferrari et al. 2006; Horstman and Kuehn 2000). Better still, density gradient centrifugation is used for ensuring removal of contaminants such as pili, flagella, other protein aggregates, and so on from the OMV preparation (Bauman and Kuehn 2006; Lee et al. 2007). On the other hand, gel filtration chromatography may be the method of choice for getting a highly purified OMV preparation. A Sephacryl S500 column was used for purification of OMVs from *Neisseria meningitidis* (Post et al. 2005). Lee et al. (2007) used a slightly modified method using two sequential steps to get a highly purified preparation of OMVs from *Escheriachia coli* DH5α cells. In the first step, bacterial cells and cell debris were removed from the culture supernatant by using differential centrifugation followed by filtration through a membrane filter (0.45 μm), precentrifugation at 20,000 and 40,000 g to remove any large vesicle or vesicle aggregates or cell debris; and then ultracentrifugation at 150,000 g. The enriched OMVs, in the second step, were further purified using density gradients to remove any remaining contaminants. Subsequent electron microscopy showed the purity of OMV preparation, the OMVs having a size between 20 – 40 nm (Lee et al. 2007).

2.4.2 *Protein Separation*

Different methods have been tried and used for separation of proteins for their subsequent analysis by mass spectrometry to get the protein profile or proteome of the OMVs. These methods include two-dimensional gel electrophoresis (2-DE) and a combination of one dimensional (1-D) SDS-PAGE (Fig. 2.6) and liquid chromatography (LC). The purified OMVs are subjected to either 2-DE or (1-DE) SDS-PAGE followed by enzymatic digestion and LC to separate the peptides. Although 2-DE is a powerful tool for protein separation, it has its limitations, particularly in the case of membrane proteins. These proteins have very poor solubility in the nondetergent isoelectric focusing buffer that causes precipitation of the proteins at their isoelectric points. In addition, the high molecular weight, basic, or hydrophobic proteins are not properly resolved by 2-DE (Post et al. 2005; Wu and Yates 2003). The other method (1-D) SDS-PAGE can separate proteins more efficiently; its limitation, particularly in high-throughput mass analysis, is greater complexity of the proteins in each gel fraction. However, this problem can be overcome by subsequent use of LC to separate the extracted peptides based on their hydrophobicity (Nevot et al. 2006; Post et al. 2005). Lee et al. (2007) argued that because the molecular weights of vesicular proteins are quite different and the OMVs also contain less abundant proteins, they used (1-D) SDS-PAGE and then cut the gel into five slices of equal size, and subjected them to trypsin digestion to extract the peptides and got better results.

Mass spectrometric analysis of the extracted peptides from the native OMVs of *E. coli* initially identified 2,606 and 2,816 proteins with high- confidence peptide sequences. In order to eliminate the peptides shared by multiple proteins, the authors used the protein hit score (PHS) method for reliable protein identification

Fig. 2.6 Protein profile of the *OMV, OM,* and *IM* of *V. cholerae*. Proteins were separated by 12 % SDS-PAGE and visualized by silver staining; Lane 1: Molecular weight (kDa) markers, Lane 2: *OMV*, Lane 3: *OM*, Lane 4: *IM*. [From (Chatterjee and Chaudhuri 2011)]

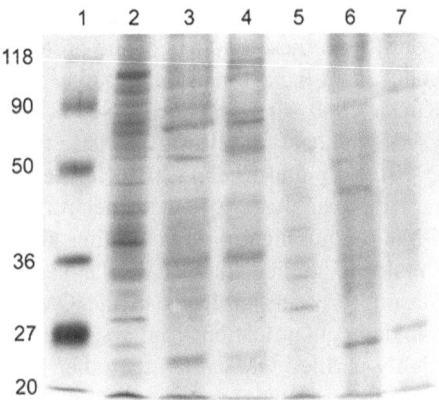

(Lee et al. 2008; Park et al. 2006). The analyses showed that proteins with PHS > 1 were identified by multiple peptides that are unique and shared with only a few proteins. Furthermore, rigorous screening of proteins of PHS > 1 identified a total of 141 proteins, including 127 previously unknown vesicular proteins, with high confidence and reproducibility (Lee et al. 2007).

Proteomic analysis of several Gram-negative bacterial OMVs defined more than 200 vesicular proteins (Bauman and Kuehn 2006; Berlanda Scorza et al. 2008; Lee et al. 2008; Post et al. 2005). When these proteins were classified into protein families based on their sequence homology and function, several families were found to be common in OMVs derived from several species of Gram-negative bacteria: (1) the abundant OMPs, Porins (Omps, PorA, PorB and OprF), which were found in most OMVs; (2) the murein hydrolases (Met and SLT), which are responsible for the hydrolysis of certain cell wall glycopeptides, peptidoglycans in particular; (3) the multidrug efflux pumps (Mtr, Mex, and TolC), which are known to release toxic compounds (Kobayashi et al. 2000); (4) the ABC transporters (LamB and FadL); (5) the protease/chaperone proteins (DegQ/SurA) and (6) the motility proteins related to fimbriae/pili (FliC/PilQ) were found in OMVs from different strains. On the other hand, the virulence factors including hemolysin, IgA protease, and macrophage infectivity potentiator were also identified in OMVs from pathogenic strains.

Proteomic analyses of OMVs have brought out some findings that may give rise to controversy. Protein profiles of OMVs and detergent-treated OMVs (DOMVs) revealed the presence of cytoplasmic proteins as well (Ferrari et al. 2006; Henry et al. 2004; Lee et al. 2007; Molloy et al. 2000; Wei et al. 2006; Xu et al. 2006). Among the cytoplasmic proteins found in the protein profiles of OMVs of different bacteria, the proteins EF-Tu, DnaK, GroEL, and two ribosomal proteins, S1 and L7/12 (which are generally highly abundant proteins), have also been detected from cell supernatants or OM fractions (Ferrari et al. 2006). It has been suggested that the transcriptional and ribosomal proteins may be sorted into the OMVs during the informational process (Dorward and Garon 1989; Dorward et al. 1989;

Ferrari et al. 2006; Kadurugamuwa and Beveridge 1995; Yaron et al. 2000). Contrary to these findings of the proteomic analyses of OMVs, many researchers believe that the cytoplasmic proteins are excluded from the OMVs (Horstman and Kuehn 2000) and that the presence of the cytoplasmic proteins, if any, in the OMVs indicates the lysis of bacteria from which they originated (Kulp and Kuehn 2010). Kulp and Kuehn (2010) further observed that in order to use the power of proteomic analyses by Mass Spectrometry to deduce the origins or biogenesis of OMVs from their protein profiles, only very carefully purified native OMVs should be studied. In fact, Berlanda Scorza et al. (2008) did not find the cytoplasmic proteins or IMPs in the OMVs derived from a log phase culture and avoided contamination from lysed cells. The authors took great care to examine highly purified OMVs. On the other hand, Lee et al. (2007) analyzed OMV proteins from a stationary phase wild- type culture of *E. coli* cells and found the presence of cytoplasmic materials, which also could be interpreted as resulting from lysis of a fraction of the bacterial cells in the resting phase. Could it be that under certain growth conditions (not yet known) the regulation of protein synthesis in the bacterial cells loses its control, leading to excess production of some cytoplasmic proteins, which the bacteria try to get rid of by secretion through OMVs? Additional studies taking great care to eliminate bacterial lysis and adopting a very stringent method of purification of OMVs might resolve the issue.

References

Allan ND, Beveridge TJ (2003) Gentamicin delivery to *Burkholderia cepacia* group IIIa strains via membrane vesicles from *Pseudomonas aeruginosa* PAO1. Antimicrob Agents Chemother 47:2962–2965

Allan ND, Kooi C, Sokol PA, Beveridge TJ (2003) Putative virulence factors are released in association with membrane vesicles from *Burkholderia cepacia*. Can J Microbiol 49:613–624

Balsalobre C, Silvan JM, Berglund S, Mizunoe Y, Uhlin BE, Wai SN (2006) Release of the type I secreted alpha-haemolysin via outer membrane vesicles from *Escherichia coli*. Mol Microbiol 59:99–112

Bauman SJ, Kuehn MJ (2006) Purification of outer membrane vesicles from *Pseudomonas aeruginosa* and their activation of an IL-8 response. Microbes Infect 8:2400–2408

Bergman MA, Cummings LA, Barrett SL, Smith KD, Lara JC, Aderem A, Cookson BT (2005) CD4+ T cells and toll-like receptors recognize *Salmonella* antigens expressed in bacterial surface organelles. Infect Immun 73:1350–1356

Berlanda Scorza F, Doro F, Rodriguez-Ortega MJ, Stella M, Liberatori S et al (2008) Proteomics characterization of outer membrane vesicles from the extraintestinal pathogenic *Escherichia coli* DeltatolR IHE3034 mutant. Mol Cell Proteomics 7:473–485

Beveridge TJ (1999) Structures of gram-negative cell walls and their derived membrane vesicles. J Bacteriol 181:4725–4733

Bjerre A, Brusletto B, Rosenqvist E, Namork E, Kierulf P et al (2000) Cellular activating properties and morphology of membrane-bound and purified meningococcal lipopolysaccharide. J Endotoxin Res 6:437–445

Boardman BK, Meehan BM, Fullner Satchell KJ (2007) Growth phase regulation of *Vibrio cholerae* RTX toxin export. J Bacteriol 189:1827–1835

Bomberger JM, Maceachran DP, Coutermarsh BA, Ye S, O'Toole GA, Stanton BA (2009) Long-distance delivery of bacterial virulence factors by *Pseudomonas aeruginosa* outer membrane vesicles. PLoS Pathog 5:e1000382

Chatterjee D, Chaudhuri K (2011) Association of cholera toxin with Vibrio cholerae outer membrane vesicles which are internalized by human intestinal epithelial cells. FEBS Lett 585:1357–1362

Chatterjee SN, Das J (1966) Secretory activity of *Vibrio cholerae* as evidenced by electron microscopy. In: Uyeda (ed) Electron microscopy, Maruzen Co. Ltd, Tokyo

Chatterjee SN, Das J (1967) Electron microscopic observations on the excretion of cell-wall material by *Vibrio cholerae*. J Gen Microbiol 49:1–11

Chatterjee SN, Chaudhuri K (2003) Lipopolysaccharides of *Vibrio cholerae*. I. Physical and chemical characterization. Biochim Biophys Acta 1639:65–79

Chaudhuri K, Chatterjee SN (2009) Cholera toxins. Springer, Heidelberg

Chi B, Qi M, Kuramitsu HK (2003) Role of dentilisin in *Treponema denticola* epithelial cell layer penetration. Res Microbiol 154:637–643

Ciofu O, Beveridge TJ, Kadurugamuwa J, Walther-Rasmussen J, Hoiby N (2000) Chromosomal beta-lactamase is packaged into membrane vesicles and secreted from *Pseudomonas aeruginosa*. J Antimicrob Chemother 45:9–13

Cota-Gomez A, Vasil AI, Kadurugamuwa J, Beveridge TJ, Schweizer HP, Vasil ML (1997) PlcR1 and PlcR2 are putative calcium-binding proteins required for secretion of the hemolytic phospholipase C of *Pseudomonas aeruginosa*. Infect Immun 65:2904–2913

Deatherage BL, Lara JC, Bergsbaken T, Rassoulian Barrett SL, Lara S, Cookson BT (2009) Biogenesis of bacterial membrane vesicles. Mol Microbiol 72:1395–1407

Dorward DW, Garon CF (1989) DNA-binding proteins in cells and membrane blebs of *Neisseria gonorrhoeae*. J Bacteriol 171:4196–4201

Dorward DW, Garon CF, Judd RC (1989) Export and intercellular transfer of DNA via membrane blebs of *Neisseria gonorrhoeae*. J Bacteriol 171:2499–2505

Dorward DW, Schwan TG, Garon CF (1991) Immune capture and detection of *Borrelia burgdorferi* antigens in urine, blood, or tissues from infected ticks, mice, dogs, and humans. J Clin Microbiol 29:1162–1170

Duncan L, Yoshioka M, Chandad F, Grenier D (2004) Loss of lipopolysaccharide receptor CD14 from the surface of human macrophage-like cells mediated by *Porphyromonas gingivalis* outer membrane vesicles. Microb Pathog 36:319–325

Dutta S, Iida K, Takade A, Meno Y, Nair GB, Yoshida S (2004) Release of Shiga toxin by membrane vesicles in *Shigella dysenteriae* serotype 1 strains and in vitro effects of antimicrobials on toxin production and release. Microbiol Immunol 48:965–969

Ellis TN, Kuehn MJ (2010) Virulence and immunomodulatory roles of bacterial outer membrane vesicles. Microbiol Mol Biol Rev 74:81–94

Fernandez-Moreira E, Helbig JH, Swanson MS (2006) Membrane vesicles shed by *Legionella pneumophila* inhibit fusion of phagosomes with lysosomes. Infect Immun 74:3285–3295

Ferrari G, Garaguso I, Adu-Bobie J, Doro F, Taddei AR et al (2006) Outer membrane vesicles from group B *Neisseria meningitidis* delta gna33 mutant: proteomic and immunological comparison with detergent-derived outer membrane vesicles. Proteomics 6:1856–1866

Forsberg CW, Beveridge TJ, Hellstrom A (1981) Cellulase and xylanase release from *Bacteroides succinogenes* and its importance in the rumen environment. Appl Environ Microbiol 42:886–896

Galka F, Wai SN, Kusch H, Engelmann S, Hecker M et al. (2008) Proteomic characterization of the whole secretome of *Legionella pneumophila* and functional analysis of outer membrane vesicles. Infect Immun 76:1825–1836

Gamazo C, Moriyon I (1987) Release of outer membrane fragments by exponentially growing *Brucella melitensis* cells. Infect Immun 55:609–615

Goulhen F, Hafezi A, Uitto VJ, Hinode D, Nakamura R, Grenier D, Mayrand D (1998) Subcellular localization and cytotoxic activity of the GroEL-like protein isolated from *Actinobacillus actinomycetemcomitans*. Infect Immun 66:5307–5313

Grenier D, Mayrand D (1987) Functional characterization of extracellular vesicles produced by *Bacteroides gingivalis*. Infect Immun 55:111–117

Grenier D (1992) Inactivation of human serum bactericidal activity by a trypsin like protease isolated from *Porphyromonas gingivalis*. Infect Immun 60:1854–1857

Guo L, Fatig RO 3rd, Orr GL, Schafer BW, Strickland JA et al (1999) *Photorhabdus luminescens* W-14 insecticidal activity consists of at least two similar but distinct proteins. Purification and characterization of toxin A and toxin B. J Biol Chem 274:9836–9842

Henry T, Pommier S, Journet L, Bernadac A, Gorvel JP, Lloubes R (2004) Improved methods for producing outer membrane vesicles in gram-negative bacteria. Res Microbiol 155:437–446

Hong GE, Kim DG, Park EM, Nam BH, Kim YO, Kong IS (2009) Identification of *Vibrio anguillarum* outer membrane vesicles related to immunostimulation in the Japanese flounder, *Paralichthys olivaceus*. Biosci Biotechnol Biochem 73:437–439

Horstman AL, Kuehn MJ (2000) Enterotoxigenic *Escherichia coli* secretes active heat-labile enterotoxin via outer membrane vesicles. J Biol Chem 275:12489–12496

Hozbor D, Rodriguez ME, Fernandez J, Lagares A, Guiso N, Yantorno O (1999) Release of outer membrane vesicles from *Bordetella pertussis*. Curr Microbiol 38:273–278

Hynes SO, Keenan JI, Ferris JA, Annuk H, Moran AP (2005) Lewis epitopes on outer membrane vesicles of relevance to *Helicobacter pylori* pathogenesis. Helicobacter 10:146–156

Kadurugamuwa JL, Beveridge TJ (1995) Virulence factors are released from *Pseudomonas aeruginosa* in association with membrane vesicles during normal growth and exposure to gentamicin: a novel mechanism of enzyme secretion. J Bacteriol 177:3998–4008

Kadurugamuwa JL, Beveridge TJ (1998) Delivery of the non-membrane-permeative antibiotic gentamicin into mammalian cells by using *Shigella flexneri* membrane vesicles. Antimicrob Agents Chemother 42:1476–1483

Kahnt J, Aguiluz K, Koch J, Treuner-Lange A, Konovalova A et al (2010) Profiling the outer membrane proteome during growth and development of the social bacterium *Myxococcus xanthus* by selective biotinylation and analyses of outer membrane vesicles. J Proteome Res 9:5197–5208

Kamaguchi A, Nakayama K, Ichiyama S, Nakamura R, Watanabe T et al. (2003) Effect of *Porphyromonas gingivalis* vesicles on coaggregation of *Staphylococcus aureus* to oral microorganisms. Curr Microbiol 47:485–491

Karched M, Ihalin R, Eneslatt K, Zhong D, Oscarsson J et al (2008) Vesicle-independent extracellular release of a proinflammatory outer membrane lipoprotein in free-soluble form. BMC Microbiol 8:18

Kato S, Kowashi Y, Demuth DR (2002) Outer membrane-like vesicles secreted by *Actinobacillus actinomycetemcomitans* are enriched in leukotoxin. Microb Pathog 32:1–13

Keenan J, Day T, Neal S, Cook B, Perez–Perez G, Allardyce R, Bagshaw P (2000) A role for the bacterial outer membrane in the pathogenesis of *Helicobacter pylori* infection. FEMS Microbiol Lett 182:259–264

Khandelwal P, Banerjee-Bhatnagar N (2003) Insecticidal activity associated with the outer membrane vesicles of *Xenorhabdus nematophilus*. Appl Environ Microbiol 69:2032–2037

Kobayashi H, Uematsu K, Hirayama H, Horikoshi K (2000) Novel toluene elimination system in a toluene-tolerant microorganism. J Bacteriol 182:6451–6455

Kolling GL, Matthews KR (1999) Export of virulence genes and Shiga toxin by membrane vesicles of *Escherichia coli* O157:H7. Appl Environ Microbiol 65:1843–1848

Kouokam JC, Wai SN, Fallman M, Dobrindt U, Hacker J, Uhlin BE (2006) Active cytotoxic necrotizing factor 1 associated with outer membrane vesicles from uropathogenic *Escherichia coli*. Infect Immun 74:2022–2030

Kuehn MJ, Kesty NC (2005) Bacterial outer membrane vesicles and the host-pathogen interaction. Genes Dev 19:2645–2655

Kulp A, Kuehn MJ (2010) Biological functions and biogenesis of secreted bacterial outer membrane vesicles. Annu Rev Microbiol 64:163–184

Kwon SO, Gho YS, Lee JC, Kim SI (2009) Proteome analysis of outer membrane vesicles from a clinical *Acinetobacter baumannii* isolate. FEMS Microbiol Lett 297:150–156

Lee EY, Bang JY, Park GW, Choi DS, Kang JS et al (2007) Global proteomic profiling of native outer membrane vesicles derived from *Escherichia coli*. Proteomics 7:3143–3153

Lee EY, Choi DS, Kim KP, Gho YS (2008) Proteomics in gram-negative bacterial outer membrane vesicles. Mass Spectrom Rev 27:535–555

Li Z, Clarke AJ, Beveridge TJ (1998) Gram-negative bacteria produce membrane vesicles which are capable of killing other bacteria. J Bacteriol 180:5478–5483

Lindmark B, Rompikuntal PK, Vaitkevicius K, Song T, Mizunoe Y et al (2009) Outer membrane vesicle-mediated release of cytolethal distending toxin (CDT) from *Campylobacter jejuni*. BMC Microbiol 9:220

MacEachran DP, Ye S, Bomberger JM, Hogan DA, Swiatecka-Urban A, Stanton BA, O'Toole GA (2007) The *Pseudomonas aeruginosa* secreted protein PA2934 decreases apical membrane expression of the cystic fibrosis transmembrane conductance regulator. Infect Immun 75:3902–3912

Mashburn-Warren LM, Whiteley M (2006) Special delivery: vesicle trafficking in prokaryotes. Mol Microbiol 61:839–846

Mashburn LM, Whiteley M (2005) Membrane vesicles traffic signals and facilitate group activities in a prokaryote. Nature 437:422–425

Masignani V, Balducci E, Di Marcello F, Savino S, Serruto D et al (2003) NarE: a novel ADP-ribosyltransferase from *Neisseria meningitidis*. Mol Microbiol 50:1055–1067

McBroom AJ, Kuehn MJ (2005) Outer membrane vesicles In: III RC (ed) EcoSal—*Escherichia coli* and *Salmonella* : cellular and molecular biology. American Society for Microbiology Press, Washington

Molloy MP, Herbert BR, Slade MB, Rabilloud T, Nouwens AS, Williams KL, Gooley AA (2000) Proteomic analysis of the *Escherichia coli* outer membrane. Eur J Biochem 267:2871–2881

Negrete-Abascal E, Garcia RM, Reyes ME, Godinez D, de la Garza M (2000) Membrane vesicles released by *Actinobacillus pleuropneumoniae* contain proteases and Apx toxins. FEMS Microbiol Lett 191:109–113

Nevot M, Deroncele V, Messner P, Guinea J, Mercade E (2006) Characterization of outer membrane vesicles released by the psychrotolerant bacterium *Pseudoalteromonas antarctica* NF3. Environ Microbiol 8:1523–1533

Park GW, Kwon KH, Kim JY, Lee JH, Yun SH et al (2006) Human plasma proteome analysis by reversed sequence database search and molecular weight correlation based on a bacterial proteome analysis. Proteomics 6:1121–1132

Patrick S, McKenna JP, O'Hagan S, Dermott E (1996) A comparison of the haemagglutinating and enzymic activities of *Bacteroides fragilis* whole cells and outer membrane vesicles. Microb Pathog 20:191–202

Post DM, Zhang D, Eastvold JS, Teghanemt A, Gibson BW, Weiss JP (2005) Biochemical and functional characterization of membrane blebs purified from *Neisseria meningitidis* serogroup B. J Biol Chem 280:38383–38394

Rosen G, Naor R, Rahamim E, Yishai R, Sela MN (1995) Proteases of *Treponema denticola* outer sheath and extracellular vesicles. Infect Immun 63:3973–3979

Schlichting E, Lyberg T, Solberg O, Andersen BM (1993) Endotoxin liberation from *Neisseria meningitidis* correlates to their ability to induce procoagulant and fibrinolytic factors in human monocytes. Scand J Infect Dis 25:585–594

Shoberg RJ, Thomas DD (1993) Specific adherence of *Borrelia burgdorferi* extracellular vesicles to human endothelial cells in culture. Infect Immun 61:3892–3900

Sidhu VK, Vorholter FJ, Niehaus K, Watt SA (2008) Analysis of outer membrane vesicle associated proteins isolated from the plant pathogenic bacterium *Xanthomonas campestris* pv. campestris. BMC Microbiol 8:87

Tan TT, Morgelin M, Forsgren A, Riesbeck K (2007) *Hemophilus influenzae* survival during complement-mediated attacks is promoted by *Moraxella catarrhalis* outer membrane vesicles. J Infect Dis 195:1661–1670

Vipond C, Suker J, Jones C, Tang C, Feavers IM, Wheeler JX (2006) Proteomic analysis of a meningococcal outer membrane vesicle vaccine prepared from the group B strain NZ98/254. Proteomics 6:3400–3413

Wai SN, Lindmark B, Soderblom T, Takade A, Westermark M et al (2003) Vesicle-mediated export and assembly of pore-forming oligomers of the enterobacterial ClyA cytotoxin. Cell 115:25–35

Wei C, Yang J, Zhu J, Zhang X, Leng W et al (2006) Comprehensive proteomic analysis of *Shigella flexneri* 2a membrane proteins. J Proteome Res 5:1860–1865

Wu CC, Yates JR 3rd (2003) The application of mass spectrometry to membrane proteomics. Nat Biotechnol 21:262–267

Xia XX, Han MJ, Lee SY, Yoo JS (2008) Comparison of the extracellular proteomes of *Escherichia coli* B and K-12 strains during high cell density cultivation. Proteomics 8:2089–2103

Xu C, Lin X, Ren H, Zhang Y, Wang S, Peng X (2006) Analysis of outer membrane proteome of *Escherichia coli* related to resistance to ampicillin and tetracycline. Proteomics 6:462–473

Yaron S, Kolling GL, Simon L, Matthews KR (2000) Vesicle-mediated transfer of virulence genes from *Escherichia coli* O157:H7 to other enteric bacteria. Appl Environ Microbiol 66:4414–4420

Yokoyama K, Horii T, Yamashino T, Hashikawa S, Barua S et al (2000) Production of shiga toxin by *Escherichia coli* measured with reference to the membrane vesicle-associated toxins. FEMS Microbiol Lett 192:139–144

Yoon H, Ansong C, Adkins JN, Heffron F (2011) Discovery of *Salmonella* virulence factors translocated via outer membrane vesicles to murine macrophages. Infect Immun 79:2182–2192

Zhu W, Thomas CE, Chen CJ, Van Dam CN, Johnston RE, Davis NL, Sparling PF (2005) Comparison of immune responses to gonococcal PorB delivered as outer membrane vesicles, recombinant protein, or Venezuelan equine encephalitis virus replicon particles. Infect Immun 73:7558–7568

Chapter 3
Factors Affecting Production of Outer Membrane Vesicles

Abstract The production of outer membrane vesicles (OMVs) by Gram-negative bacteria is influenced by many different factors. Pathogenic bacteria produce more OMVs than the non-pathogenic ones and OMVs are also produced within the infected hosts. The amount of OMVs produced under different growth conditions varies, and the structure of LPS on the outer membrane significantly influences OMV production. Bacteria treated with antibiotics such as gentamicin produce numerous OMVs that are different from the native OMVs in structure and chemical composition. Similarly, bacteria under stress also produce more OMVs. Thus the OMVs are produced to favor the growth and survival of the parent bacteria under challenging conditions.

Keywords Natural and unnatural OMVs · Pathogenic and non-pathogenic bacteria · Growth conditions · LPS structure · Infected host · Antibiotic treatment · Stress response

3.1 Natural and Unnatural OMVs

Outer membrane vesicles (OMVs) are produced by Gram-negative bacteria, in general, during their active growth and not when they undergo lysis and death (Chatterjee and Das 1966, 1967). In fact, the OMVs have been found to contain newly synthesized proteins and are produced without concomitant bacterial lysis (Ellis and Kuehn 2010; McBroom et al. 2006; Mug-Opstelten and Witholt 1978; Zhou et al. 1998). However, OMVs or OMV-like particles have also been produced by certain drastic treatments (detergent, antibiotics etc.) or during abnormal or unbalanced growth in nutritionally deficient media or after artificial treatment (sonication, etc.) of bacteria. These OMVs or better OMV-like particles

S. N. Chatterjee and K. Chaudhuri, *Outer Membrane Vesicles of Bacteria*,
SpringerBriefs in Microbiology, DOI: 10.1007/978-3-642-30526-9_3,
© The Author(s) 2012

are compositionally different from naturally produced OMVs and they mostly contain materials leaking from bacteria undergoing lysis. It is proposed that these OMVs be termed unnatural OMVs vis-à-vis the natural ones formed by bacteria growing normally in a nutritionally rich culture medium or within the infected host. Genetic studies by generating transposon insertion mutants of *Escherichia coli* revealed that vesiculation or OMV production is not a consequence of bacterial lysis or disintegration of the bacterial envelope and cannot be correlated with membrane instability, and that vesiculation is a process important in the growth of Gram-negative bacteria (McBroom et al. 2006). Accordingly, the unnatural OMVs are discussed briefly as and when required for relevance only.

3.2 Pathogenic and Nonpathogenic Bacteria

OMVs are produced by both pathogenic and nonpathogenic species of Gram-negative bacteria (Beveridge 1999; Chatterjee and Das 1966, 1967; Kadurugamuwa and Beveridge 1997; Li et al. 1998; Mayrand and Grenier 1989). The different bacterial species that have already been demonstrated to release OMVs include *E. coli* (Gankema et al. 1980; Hoekstra et al. 1976), *Shigella spp.* (Dutta et al. 2004; Kadurugamuwa and Beveridge 1999), *Neisseria spp.* (Devoe and Gilchrist 1973; Dorward and Garon 1989; Dorward et al. 1989), *Pseudomonas aeruginosa* (Kadurugamuwa and Beveridge 1995), *Vibrio spp.* (Chatterjee and Das 1966, 1967; Iwanaga and Naito 1979, 1980; Kondo et al. 1993), *Helicobacter pylori* (Fiocca et al. 1999), *Salmonella spp.* (Vesy et al. 2000; Wai et al. 2003), *Brucella melitensis* (Gamazo and Moriyon 1987), *Bacteroides* (including *Porphyromonas*) *spp.* (Grenier and Mayrand 1987; Mayrand and Holt 1988; Zhou et al. 1998), *Borrelia burgdorferi* (Shoberg and Thomas 1993), and *Actinobacillus actinomycetemcomitans* (Nowotny et al. 1982). In general, pathogenic bacteria produce more vesicles than the corresponding nonpathogenic ones (Lai et al. 1981; Wai et al. 1995). In fact, Enterotoxigenic *E. coli* cells were found to produce tenfold more vesicles than their corresponding nonpathogenic ones (Horstman and Kuehn 2002). Similarly the pathogenic leukotoxic strains of *A. actinomycetem-comitans* were shown to produce 25-fold more vesicles than their corresponding nonpathogenic ones (Lai et al. 1981). *E. coli* strains bearing a mutation in *hns*, a virulence regulatory factor, produced threefold more vesicles (Horstman and Kuehn 2002). This evidence gives credence to the idea that vesicle production is utilized by the pathogenic bacteria to disseminate virulence factors and gain better survival in the host. Similarly, nonpathogenic bacteria can also take recourse to vesicle production for improving survival by releasing different toxic compounds, such as toluene, and by aiding in the release or removal of the attacking phages (Kobayashi et al. 2000; Loeb 1974; Loeb and Kilner 1978).

Fig. 3.1 A dividing
V. cholerae cell, thin-
sectioned, stained with
potassium permanganate and
electron micrographed.
Saclike structures (*SS*) or
OMVs formed by the bulged-
out cell wall portion are
presumably ready to be
pinched off. Bar represents
0.1μm. From (Chatterjee and
Das 1967)

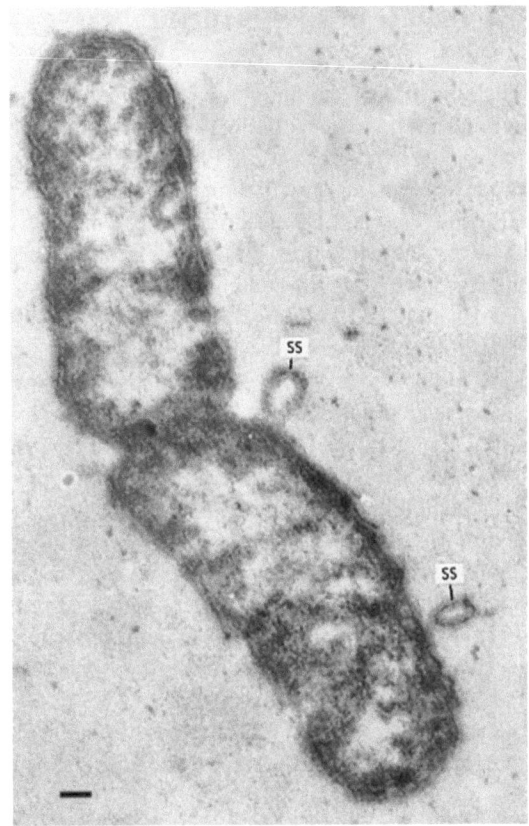

3.3 Bacterial Growth Conditions

Gram-negative bacteria growing (1) in liquid culture (Chatterjee and Das 1966, 1967; Wai et al. 1995), (2) on solid growth media (Tetz et al. 1990; Unal et al. 2010), (3) in biofilms (Schooling and Beveridge 2006; Unal et al. 2010; Yonezawa et al. 2009), and (4) within hosts (Avakian et al. 1972; Fiocca et al. 1999; Galka et al. 2008; Stephens et al. 1982) produce OMVs for various purposes. OMVs are produced more actively during the exponential growth phase (Chatterjee and Das 1966, 1967) and during division at the site of division and elsewhere (Fig. 3.1) (Chatterjee and Das 1967; Deatherage et al. 2009; Kuehn and Kesty 2005). During the resting or stationary phase of growth, OMVs are not produced (Chatterjee and Das 1967) or are produced in a very limited way (Bauman and Kuehn 2006; Hoekstra et al. 1976). It may be that during the resting phase of bacterial population growth, some individual bacteria may still remain in the logarithmic phase and produce OMVs. This evidence is again consistent with the idea that OMV production is an inherent property of the bacteria and is produced for some purposes in favor of the actively growing parent bacteria.

3.4 Impact of LPS Structure

The structure of the antigenic LPS chains projecting outward from the bacterial surface has a significant effect on the formation of OMVs. There are bacterial spp. (e.g., *P. aeruginosa* strain PAO1) that express two types of LPS chains and produce vesicles enriched in the highly charged and longer "B-band" form (Beveridge et al. 1997; Kadurugamuwa and Beveridge 1995; Nguyen et al. 2003). Such enrichment happens probably because charge–charge repulsion takes place in the regions of the OM containing adjacent B-band LPS molecules leading to local deformation and budding of the OM (Kadurugamuwa and Beveridge 1996; Li et al. 1996). Thus, *P. aeruginosa* strain PAO1 was shown to produce more B-band LPS and an increase in vesiculation when grown under oxygen stress conditions (Sabra et al. 2003). Mutants of the *Salmonella* and *P. aeruginosa* strains having no LPS O-antigen side chain produced more OMVs (Meadow et al. 1978; Smit et al. 1975). On the other hand, mutations in the core region of LPS were shown to be associated with decreased expression of outer membrane proteins (OMPs) (Ames et al. 1974; Schnaitman and Klena 1993; Smit et al. 1975). It was thus interpreted that the vesiculation phenotypes produced in LPS core mutants were the result of alteration in OMP composition and hence were the indirect effects (Meadow et al. 1978) of core mutation. A typical defense strategy of bacteria is to alter expression of LPS O-antigen to evade the host response (Lerouge and Vanderleyden 2002; Pier 2000). The presence and type of LPS O-antigen as well as the indirect effect of oxygen stress on the LPS structure may influence the physical ability of the membrane to bulge and initiate the formation of vesicles.

3.5 Vesicle Production Within the Infected Host

Gram-negative bacteria have been shown to produce OMVs while present within the host system (Ellis and Kuehn 2010) and in a variety of environments. Presence of antibiotics, availability of iron, LPS phenotype switching, and oxygen stress are some of the factors the bacteria face within the infected host and these conditions influence vesicle production there (Kuehn and Kesty 2005). Perhaps the earliest evidence in this respect was produced by Avakyan and co-workers by electron microscopy of the biopsied sections of small intestine mucosa in patients with cholera (Avakian et al. 1972) and then by Halhoul and Colvin in their study on the ultrastructure of plaque attached to human gingiva (Halhoul and Colvin 1975). However, both the composition and production of vesicles within the host systems depend on the environmental factors that the bacteria face within the host.

H. pylori are known to colonize the stomach and cause peptic ulcer and even cancer. These colonizing bacteria were shown by electron microscopy to produce vesicles that bind to gastric cells (Fiocca et al. 1999; Keenan et al. 2000; Keenan and Allardyce 2000). These bacteria experience different levels of iron within the host. These vesicles contained the vacuolating cytotoxin, VacA, and they were very similar to those produced by *H. pylori* in vitro (Fiocca et al. 1999; Keenan et al. 2000). It was shown that the vesicles derived from *H. pylori* were in contact with the intestinal epithelial cells (Fiocca et al. 1999; Heczko et al. 2000; Keenan et al. 2000). However, growth of these bacteria in iron-limiting conditions reduces VacA and increases the concentration of proteases within the vesicles while maintaining at the same time the vesicle production level (Keenan and Allardyce 2000). The OMVs from a highly vesiculating strain of *Neisseria meningitidis* causing a fatal septic infection in humans were the causative factors of a high level of endotoxins in the cells (Namork and Brandtzaeg 2002). Mouse fibroblast cells infected with *Chlamydia trachomatis* or *Chlamydia psittaci* were shown to contain vesicles (Stirling and Richmond 1980). When the *B. burgdorferi* isolate (causative factor of Lyme disease) was incubated with human skin for about 24 h, vesicles could be detected after the organisms invaded the dermis (Beermann et al. 2000).

Different fluids isolated from the infected hosts also contained vesicles. This indicated that the vesicles could migrate to sites at a distance from the point of infection. *N. meningitidis* along with the vesicles they released were found in the cerebrospinal fluid from a patient with meningitis and blood from a patient who died from meningitis (Bjerre et al. 2000; Brandtzaeg et al. 1992; Craven et al. 1980; Namork and Brandtzaeg 2002; Stephens et al. 1982). Similarly, vesiculating *Borrelia* and free vesicles were detected in the blood and urine of *B. burgdorferi*-infected mice and also ticks (Dorward et al. 1991). Also, *Salmonella typhimurium* organisms were shown to produce vesicles when growing intracellularly and also in the in vitro culture fluid (Bergman et al. 2005; Garcia-del Portillo et al. 1997; Vesy et al. 2000).

These findings strongly support the idea that vesicle production takes place within an infected host and that the vesicles are found surrounding the parent bacteria and/or in contact with the host cells. They are also available in different fluids collected from the infected patients. The vesicle production within the infected host is thus an inherent property of the bacteria. But to what extent these vesicles are directly responsible for spreading infection within the host by acting in "self-defense" or by killing the host or other co-colonizing bacterial cells and if this is true, whether they can be targeted or selectively destroyed to save the patient remains an important subject for further studies.

3.6 Antibiotic Treatment and Vesicle Formation

Treatment of bacteria with antibiotics often leads to unnatural vesicle production. Several aspects of vesiculation were shown to be affected by antibiotic treatment and the response differed with the antibiotic. Of the different antibiotics available, gentamicin in particular has been studied in detail. *P. aeruginosa* on treatment with gentamicin produces at least three fold more vesicles and the structure of these vesicles is significantly different from those produced by the same bacteria under normal growth conditions (Kadurugamuwa and Beveridge 1998). The gentamicin-induced vesicles serve an important purpose, that is, they fuse to host cells infected by pathogenic bacteria, deliver the antibiotic, gentamicin, in the cytosol and thereby kill the intracellular bacteria (Kadurugamuwa and Beveridge 1998). The gentamicin-induced vesicles are larger in size *vis-à-vis* those produced by normally growing bacteria and contain in addition to the OM and periplasmic components, some components of the IM and the cytosol (Kadurugamuwa and Beveridge 1995). In addition, the gentamicin-induced vesicles of *P. aeruginosa* are not enriched in the B-band LPS that are found in native vesicles. Although the gentamicin-induced vesicles appear to be produced by a different mechanism, they served the purpose of elucidating the fusogenic capacity of the vesicles and their ability to interact with neighboring cells. The gentamicin-induced *P. aeruginosa* vesicles were also bacteriolytic, a mechanism that helped the bacteria in securing a niche in a competitive microbial environment (Kadurugamuwa and Beveridge 1996; Allan and Beveridge 2003). In *Shigella dysenteriae*, treatment with mito-mycin C caused production of Shiga toxin and also increased production of OMVs of greater sizes and toxicity (Dutta et al. 2004). Treatment with some other antibiotics produced different responses (Dutta et al. 2004). Treatment with the antibiotics, fosfomycin, ciprofloxacin, and norfloxacin, did not have any signifi-cant effect on vesicle production or toxicity (Kuehn and Kesty 2005).

3.7 Stress Response and OMV Production

Envelope stress is produced by several factors including impairment of protein folding in the periplasm. Increased production of OMVs is a significant one among different modes of bacterial stress response. There are several mechanisms of invoking stress responses by Gram-negative bacteria facing different stressors (Raivio 2005). The σ^E is one of the different stress response pathways and is activated by events or mutations that lead to alterations in OMP biogenesis including misfolding of proteins in the periplasm. Protein misfolding in the periplasm leads to the activation of several events: (1) the membrane-bound antisigma factor, RseA, is cleaved by the protease, DegS; (2) normally another periplasmic regulatory molecule, RseB, protects RseA from cleavage in the absence of inducing signals; (3) the degradation of RseA leads to the release of σ^E into the cytoplasm and the transcriptional activation of a set of genes that include many involved in OMP and outer-membrane biogenesis (Raivio 2005). Two of the genes, *degS* and *rseA*, code for stress signal transmitters in the σ^E stress-response pathway, and another gene, *degP*, codes for a downstream effector. Activity of σ^E is essential under both stress and nonstress conditions, and in addition to its role in monitoring and maintaining OMPs in the face of adverse conditions, σ^E plays other key physiological roles (De Las Penas et al. 1997). The Cpx is another envelope stress-response pathway that appears to maintain envelope protein folding status in the presence of adverse conditions (Raivio 2005). McBroom et al. (2006)

revealed that vesiculation levels were altered by mutation of an envelope stress-response pathway. Some disruption of genes in this pathway was known to result in low σ^E activity, whereas others caused no change or even hyperactivation of the σ^E response (Alba and Gross 2004). Interestingly, all the σ^E pathway mutants of *E. coli* obtained by McBroom et al. (2006) caused increased vesiculation. The authors proposed that impairment and hyperactivation of the σ^E pathway perhaps resulted in accumulation of materials in the cell envelope, which induced heightened vesiculation. They studied the relation between vesiculation and activation of the σ^E pathway and thought that vesiculation under stressing conditions possibly occurred via a mechanism that differed from the typical vesiculation process.

McBroom and Kuehn presented data that revealed another novel stress response mechanism of Gram-negative bacteria, the release of outer membrane vesicles (McBroom and Kuehn 2007). By using an elegant genetic method (transposon mutagenesis screen) and several *E. coli* mutants, they showed that vesicle production is not directly correlated with σ^E pathway activity; that is there are other means of regulating the vesiculation level. The significant findings of their study include: (1) vesiculation increases in response to impairment of the σ^E pathway, (2) vesiculation is regulated by the level of protein accumulation in the envelope periplasm, (3) mutations that cause increased vesiculation improve bacterial survival under stress including accumulation of toxic misfolded proteins, (4) vesiculation is a distinctly independent stress response, (5) vesiculation does not involve any significant loss of membrane integrity, and (6) the vesiculation process can act to selectively eliminate unwanted materials such as misfolded proteins in the periplasm.

Involvement of a different entity in the process of OMV formation was presented (Song et al. 2008). A small noncoding s-RNA gene, *vrrA*, was discovered in *V. cholerae* O1 strain A1552. The corresponding VrrA RNA (140 nt) was found to repress the *ompA* translation. It was shown that the expression of the *vrrA* gene required the membrane stress factor σ^E, suggesting that *vrrA* acted on *ompA* in response to periplasmic protein-folding stress. The OmpA levels were again found to correlate inversely with the number of OMVs produced and that VrrA increased OMV production comparable to loss of OmpA. VrrA was thus shown to control OMV production. It was proposed that VrrA acted as a regulator mediating σ^E related stress.

References

Alba BM, Gross CA (2004) Regulation of the *Escherichia coli* sigma-dependent envelope stress response. Mol Microbiol 52:613–619

Allan ND, Beveridge TJ (2003) Gentamicin delivery to *Burkholderia cepacia* group IIIa strains via membrane vesicles from Pseudomonas aeruginosa PAO1. Antimicrob Agents Chemother 47:2962–2965

Ames GF, Spudich EN, Nikaido H (1974) Protein composition of the outer membrane of *Salmonella typhimurium*: effect of lipopolysaccharide mutations. J Bacteriol 117:406–416

Avakian AA, Sinel'nikova MP, Pereverzev NA, Gurskii IuN (1972) Electron microscopic study of biopsied sections of small intestine mucosa in patients with cholera and characteristics of the ultrastructure of causative agents of cholera in relation to toxinogenesis. Zh Mikrobiol Epidemiol Immunobiol 49:86–92

Bauman SJ, Kuehn MJ (2006) Purification of outer membrane vesicles from *Pseudomonas aeruginosa* and their activation of an IL-8 response. Microbes Infect 8:2400–2408

Beermann C, Wunderli-Allenspach H, Groscurth P, Filgueira L (2000) Lipoproteins from *Borrelia burgdorferi* applied in liposomes and presented by dendritic cells induce CD8(+) T-lymphocytes in vitro. Cell Immunol 201:124–131

Bergman MA, Cummings LA, Barrett SL, Smith KD, Lara JC, Aderem A, Cookson BT (2005) CD4 + T cells and toll-like receptors recognize *Salmonella* antigens expressed in bacterial surface organelles. Infect Immun 73:1350–1356

Beveridge TJ (1999) Structures of gram-negative cell walls and their derived membrane vesicles. J Bacteriol 181:4725–4733

Beveridge TJ, Makin SA, Kadurugamuwa JL, Li Z (1997) Interactions between biofilms and the environment. FEMS Microbiol Rev 20:291–303

Bjerre A, Brusletto B, Rosenqvist E, Namork E, Kierulf P et al (2000) Cellular activating properties and morphology of membrane-bound and purified meningococcal lipopolysaccharide. J Endotoxin Res 6:437–445

Brandtzaeg P, Bryn K, Kierulf P, Ovstebo R, Namork E, Aase B, Jantzen E (1992) Meningococcal endotoxin in lethal septic shock plasma studied by gas chromatography, mass-spectrometry, ultracentrifugation, and electron microscopy. J Clin Invest 89:816–823

Chatterjee SN, Das J (1966) Secretory activity of *Vibrio cholerae* as evidenced by electron microscopy. In: Uyeda R (ed) Electron Microscopy 1966. Maruzen Co. Ltd, Tokyo

Chatterjee SN, Das J (1967) Electron microscopic observations on the excretion of cell-wall material by *Vibrio cholerae*. J Gen Microbiol 49:1–11

Craven DE, Peppler MS, Frasch CE, Mocca LF, McGrath PP, Washington G (1980) Adherence of isolates of *Neisseria meningitidis* from patients and carriers to human buccal epithelial cells. J Infect Dis 142:556–568

De Las Penas A, Connolly L, Gross CA (1997) SigmaE is an essential sigma factor in *Escherichia coli*. J Bacteriol 179:6862–6864

Deatherage BL, Lara JC, Bergsbaken T, Rassoulian Barrett SL, Lara S, Cookson BT (2009) Biogenesis of bacterial membrane vesicles. Mol Microbiol 72:1395–1407

Devoe IW, Gilchrist JE (1973) Release of endotoxin in the form of cell wall blebs during in vitro growth of *Neisseria meningitidis*. J Exp Med 138:1156–1167

Dorward DW, Garon CF (1989) DNA-binding proteins in cells and membrane blebs of *Neisseria gonorrhoeae*. J Bacteriol 171:4196–4201

Dorward DW, Garon CF, Judd RC (1989) Export and intercellular transfer of DNA via membrane blebs of *Neisseria gonorrhoeae*. J Bacteriol 171:2499–2505

Dorward DW, Schwan TG, Garon CF (1991) Immune capture and detection of *Borrelia burgdorferi* antigens in urine, blood, or tissues from infected ticks, mice, dogs, and humans. J Clin Microbiol 29:1162–1170

Dutta S, Iida K, Takade A, Meno Y, Nair GB, Yoshida S (2004) Release of Shiga toxin by membrane vesicles in *Shigella dysenteriae* serotype 1 strains and in vitro effects of antimicrobials on toxin production and release. Microbiol Immunol 48:965–969

Ellis TN, Kuehn MJ (2010) Virulence and immunomodulatory roles of bacterial outer membrane vesicles. Microbiol Mol Biol Rev 74:81–94

Fiocca R, Necchi V, Sommi P, Ricci V, Telford J, Cover TL, Solcia E (1999) Release of *Helicobacter pylori* vacuolating cytotoxin by both a specific secretion pathway and budding of outer membrane vesicles. Uptake of released toxin and vesicles by gastric epithelium. J Pathol 188:220–226

Galka F, Wai SN, Kusch H, Engelmann S, Hecker M et al (2008) Proteomic characterization of the whole secretome of *Legionella pneumophila* and functional analysis of outer membrane vesicles. Infect Immun 76:1825–1836

Gamazo C, Moriyon I (1987) Release of outer membrane fragments by exponentially growing *Brucella melitensis* cells. Infect Immun 55:609–615

Gankema H, Wensink J, Guinee PA, Jansen WH, Witholt B (1980) Some characteristics of the outer membrane material released by growing enterotoxigenic *Escherichia coli*. Infect Immun 29:704–713

Garcia-del Portillo F, Stein MA, Finlay BB (1997) Release of lipopolysaccharide from intracellular compartments containing *Salmonella typhimurium* to vesicles of the host epithelial cell. Infect Immun 65:24–34

Grenier D, Mayrand D (1987) Functional characterization of extracellular vesicles produced by *Bacteroides gingivalis*. Infect Immun 55:111–117

Halhoul N, Colvin JR (1975) The ultrastructure of bacterial plaque attached to the gingiva of man. Arch Oral Biol 20:115–118

Heczko U, Smith VC, Mark Meloche R, Buchan AM, Finlay BB (2000) Characteristics of *Helicobacter pylori* attachment to human primary antral epithelial cells. Microbes Infect 2:1669–1676

Hoekstra D, van der Laan JW, de Leij L, Witholt B (1976) Release of outer membrane fragments from normally growing *Escherichia coli*. Biochim Biophys Acta 455:889–899

Horstman AL, Kuehn MJ (2002) Bacterial surface association of heat-labile enterotoxin through lipopolysaccharide after secretion via the general secretory pathway. J Biol Chem 277:32538–32545

Iwanaga M, Naito T (1979) Morphological changes of *Vibrio cholerae* during toxin production. Trop Med 21:187–196

Iwanaga M, Naito T (1980) Toxin production and morphological changes of *Vibrio cholerae* in the medium for inducing pleomorphism. Trop Med 22:61–68

Kadurugamuwa JL, Beveridge TJ (1995) Virulence factors are released from *Pseudomonas aeruginosa* in association with membrane vesicles during normal growth and exposure to gentamicin: a novel mechanism of enzyme secretion. J Bacteriol 177:3998–4008

Kadurugamuwa JL, Beveridge TJ (1996) Bacteriolytic effect of membrane vesicles from *Pseudomonas aeruginosa* on other bacteria including pathogens: conceptually new antibiotics. J Bacteriol 178:2767–2774

Kadurugamuwa JL, Beveridge TJ (1997) Natural release of virulence factors in membrane vesicles by *Pseudomonas aeruginosa* and the effect of aminoglycoside antibiotics on their release. J Antimicrob Chemother 40:615–621

Kadurugamuwa JL, Beveridge TJ (1998) Delivery of the non-membrane-permeative antibiotic gentamicin into mammalian cells by using *Shigella flexneri* membrane vesicles. Antimicrob Agents Chemother 42:1476–1483

Kadurugamuwa JL, Beveridge TJ (1999) Membrane vesicles derived from *Pseudomonas aeruginosa* and Shigella flexneri can be integrated into the surfaces of other gram-negative bacteria. Microbiology 145(Pt 8):2051–2060

Keenan JI, Allardyce RA (2000) Iron influences the expression of *Helicobacter pylori* outer membrane vesicle-associated virulence factors. Eur J Gastroenterol Hepatol 12:1267–1273

Keenan J, Day T, Neal S, Cook B, Perez-Perez G, Allardyce R, Bagshaw P (2000) A role for the bacterial outer membrane in the pathogenesis of *Helicobacter pylori* infection. FEMS Microbiol Lett 182:259–264

Kobayashi H, Uematsu K, Hirayama H, Horikoshi K (2000) Novel toluene elimination system in a toluene-tolerant microorganism. J Bacteriol 182:6451–6455

Kondo K, Takade A, Amako K (1993) Release of the outer membrane vesicles from *Vibrio cholerae* and *Vibrio parahaemolyticus*. Microbiol Immunol 37:149–152

Kuehn MJ, Kesty NC (2005) Bacterial outer membrane vesicles and the host-pathogen interaction. Genes Dev 19:2645–2655

Lai CH, Listgarten MA, Hammond BF (1981) Comparative ultrastructure of leukotoxic and non-leukotoxic strains of *Actinobacillus actinomycetemcomitans*. J Periodontal Res 16:379–389

Lerouge I, Vanderleyden J (2002) O-antigen structural variation: mechanisms and possible roles in animal/plant-microbe interactions. FEMS Microbiol Rev 26:17–47

Li Z, Clarke AJ, Beveridge TJ (1996) A major autolysin of *Pseudomonas aeruginosa*: subcellular distribution, potential role in cell growth and division and secretion in surface membrane vesicles. J Bacteriol 178:2479–2488

Li Z, Clarke AJ, Beveridge TJ (1998) Gram-negative bacteria produce membrane vesicles which are capable of killing other bacteria. J Bacteriol 180:5478–5483

Loeb MR (1974) Bacteriophage T4-mediated release of envelope components from *Escherichia coli*. J Virol 13:631–641

Loeb MR, Kilner J (1978) Release of a special fraction of the outer membrane from both growing and phage T4-infected *Escherichia coli* B. Biochim Biophys Acta 514:117–127

Mayrand D, Grenier D (1989) Biological activities of outer membrane vesicles. Can J Microbiol 35:607–613

Mayrand D, Holt SC (1988) Biology of asaccharolytic black-pigmented *Bacteroides species*. Microbiol Rev 52:134–152

McBroom AJ, Kuehn MJ (2007) Release of outer membrane vesicles by gram-negative bacteria is a novel envelope stress response. Mol Microbiol 63:545–558

McBroom AJ, Johnson AP, Vemulapalli S, Kuehn MJ (2006) Outer membrane vesicle production by *Escherichia coli* is independent of membrane instability. J Bacteriol 188:5385–5392

Meadow PM, Wells PL, Salkinoja-Salonen M, NE L (1978) The effect of lipopolysaccharide composition on the ultrastructure of *Pseudomonas aeruginosa*. J Gen Microbiol 105:23–28

Mug-Opstelten D, Witholt B (1978) Preferential release of new outer membrane fragments by exponentially growing *Escherichia coli*. Biochim Biophys Acta 508:287–295

Namork E, Brandtzaeg P (2002) Fatal meningococcal septicaemia with "blebbing" meningococcus. Lancet 360:1741

Nguyen TT, Saxena A, Beveridge TJ (2003) Effect of surface lipopolysaccharide on the nature of membrane vesicles liberated from the gram-negative bacterium *Pseudomonas aeruginosa*. J Electron Microsc (Tokyo) 52:465–469

Nowotny A, Behling UH, Hammond B, Lai CH, Listgarten M, Pham PH, Sanavi F (1982) Release of toxic microvesicles by *Actinobacillus actinomycetemcomitans*. Infect Immun 37:151–154

Pier GB (2000) Peptides, *Pseudomonas aeruginosa*, polysaccharides and lipopolysaccharides–players in the predicament of cystic fibrosis patients. Trends Microbiol 8:247–250; discussion 250–241

Raivio TL (2005) Envelope stress responses and gram-negative bacterial pathogenesis. Mol Microbiol 56:1119–1128

Sabra W, Lunsdorf H, Zeng AP (2003) Alterations in the formation of lipopolysaccharide and membrane vesicles on the surface of *Pseudomonas aeruginosa* PAO1 under oxygen stress conditions. Microbiology 149:2789–2795

Schnaitman CA, Klena JD (1993) Genetics of lipopolysaccharide biosynthesis in enteric bacteria. Microbiol Rev 57:655–682

Schooling SR, Beveridge TJ (2006) Membrane vesicles: an overlooked component of the matrices of biofilms. J Bacteriol 188:5945–5957

Shoberg RJ, Thomas DD (1993) Specific adherence of *Borrelia burgdorferi* extracellular vesicles to human endothelial cells in culture. Infect Immun 61:3892–3900

Smit J, Kamio Y, Nikaido H (1975) Outer membrane of *Salmonella typhimurium*: chemical analysis and freeze-fracture studies with lipopolysaccharide mutants. J Bacteriol 124:942–958

Song T, Mika F, Lindmark B, Liu Z, Schild S et al (2008) A new *Vibrio cholerae* sRNA modulates colonization and affects release of outer membrane vesicles. Mol Microbiol 70:100–111

Stephens DS, Edwards KM, Morris F, McGee ZA (1982) Pili and outer membrane appendages on *Neisseria meningitidis* in the cerebrospinal fluid of an infant. J Infect Dis 146:568

Stirling P, Richmond SJ (1980) Production of outer membrane blebs during chlamydial replication. FEMS Microbiol Lett 9:103–105

Tetz VV, Rybalchenko OV, Savkova GA (1990) Ultrastructural features of microbial colony organization. J Basic Microbiol 30:597–607

Unal CM, Schaar V, Riesbeck K (2010) Bacterial outer membrane vesicles in disease and preventive medicine. Semin Immunopathol 33:395–408

Vesy CJ, Kitchens RL, Wolfbauer G, Albers JJ, Munford RS (2000) Lipopolysaccharide-binding protein and phospholipid transfer protein release lipopolysaccharides from gram-negative bacterial membranes. Infect Immun 68:2410–2417

Wai SN, Takade A, Amako K (1995) The release of outer membrane vesicles from the strains of enterotoxigenic *Escherichia coli*. Microbiol Immunol 39:451–456

Wai SN, Lindmark B, Soderblom T, Takade A, Westermark M et al (2003) Vesicle-mediated export and assembly of pore-forming oligomers of the enterobacterial ClyA cytotoxin. Cell 115:25–35

Yonezawa H, Osaki T, Kurata S, Fukuda M, Kawakami H et al (2009) Outer membrane vesicles of *Helicobacter pylori* TK1402 are involved in biofilm formation. BMC Microbiol 9:197

Zhou L, Srisatjaluk R, Justus DE, Doyle RJ (1998) On the origin of membrane vesicles in gram-negative bacteria. FEMS Microbiol Lett 163:223–228

Chapter 4
Outer Membrane Vesicles as Carriers of Biomaterials

Abstract During their formation the outer membrane vesicles (OMVs) entrap, utilizing some special sorting mechanism, different materials from the periplasm and/or the outer membrane of the bacterial cell for purposes favoring the parent bacteria and against the survival of other competing organisms either in vitro or in vivo. These entrapped materials include different virulence factors, toxins and nontoxins, and other materials such as antibiotics and DNA that can either kill the competing organisms or change them genetically. The functions of these entrapped materials and the modes of their entrapment including the formation of engineered recombinant OMVs are briefly discussed.

Keywords Predatory activities · Toxins · Virulence factors · DNA · Antibiotics · Recombinant OMVs

4.1 Packaging and Predatory Activities

The OMVs derived from many Gram-negative bacteria contain different virulence factors including toxins and nontoxins. Some of these virulence factors, depending on growth conditions, are enriched in OMVs vis-à-vis bacterial periplasm. It has been found, in general, that in *E. coli* approximately 0.2–0.5% of the outer membrane and periplasmic proteins are packaged in the OMVs (Hoekstra et al. 1976; Mug-Opstelten and Witholt 1978; Kesty and Kuehn 2004). It was suggested that a sorting mechanism works during the formation of the OMVs (Kadurugamuwa and Beveridge 1995); the actual mechanism has still not been understood. The packaging of different virulence factors, toxins and nontoxins, provides some advantages: (1) OMVs may concentrate the virulence factors for focused delivery to the target cells; and (2) the luminal contents of the OMVs, the virulence factors and other chemicals, stand protected from degradation and/or recognition by hostile environmental factors (Mashburn-Warren and Whiteley 2008). It is of interest to note that the OMVs were termed "predatory" (Beveridge et al. 1997; Kadurugamuwa and Beveridge 1996; Li et al. 1998) because of their killing properties.*P. aeruginosa* OMVs contain factors critical for killing host cells as well as several antibacterial factors including murein hydrolases or autolysins (Li et al. 1996). Normally the autolysins play a role in cell-wall turnover and cell division, but the autolysin-containing OMVs are able to degrade the peptidoglycan layers of bacteria. Furthermore, *P. aeruginosa* OMVs are enriched with a 26-kDa autolysin (Li et al. 1996). It was shown that OMVs derived from many different Gram-negative bacteria were able to lyse and kill many Gram-negative as well as Gram-positive target bacteria (Li et al. 1998). Interestingly, those bacteria that had their peptidoglycan chemotype similar to that of the OMV-producing ones are most susceptible to lysis by these OMVs (Li et al. 1998). This role of autolysins also remains to be explained. One can, however, argue that the predatory OMVs provide a special advantage to the Gram-negative bacteria in that they can survive in a polymicrobial environment through lysis of nonself bacteria. Another advantage provided by the OMVs is that the lysis of target cells releases nutrients favoring enhanced growth of the parent bacteria.

4.2 Toxins

Bacteria produce and secrete different types of toxins in the extracellular medium by different mechanisms of secretion. The OMVs offer one of the means of toxin release by several bacteria. There may be several mechanisms of association of OMVs with toxins, which may vary from bacteria to bacteria. The fully formed toxin in the periplasm may get entrapped by the OMVs during their formation. These entrapped toxins may exist either in the lumen or be associated with the inner surface of the bounding membrane of the OMVs. Also they may just leach through the membrane and be associated with the LPS on the outer surface. Furthermore, the toxin may, in some cases, be directly secreted to the extracellular milieu through the IM and OM of the bacteria by some active one-step or two-step processes. Once in the extracellular medium, they get the chance to be attached to the LPS on the outer surface of the OMVs already released. However, experimental evidence for all these modes of association of toxins with the OMVs are yet to be established, although the interaction of some vesicle containing toxins with the host cell membrane has already got significant experimental support. Some examples of OMV-associated toxins include: (1) Shiga toxin from *Shigella dysenteriae* serotype 1 (Kolling and Matthews 1999; Yokoyama et al. 2000); (2) LT from enterotoxigenic *E. coli* (Horstman and Kuehn 2002); (3) leukotoxin from *Actinobacillus actinomycetemcomitans* (Kato et al. 2002); (4) Apx toxin from *Actinobacillus pleuropneumoniae* (Negrete-Abascal et al. 2000);(5) VacA from *Helicobacter pylori* (Balsalobre et al. 2006; Berlanda Scorza et al. 2008);

(6) RTX toxin (Balsalobre et al. 2006), Cytolysin A (ClyA) from *E. coli* (Wai et al. 2003); and also (7) CT (Chatterjee and Chaudhuri 2011) from *V. cholerae*, and so on. In some cases, vesicle-associated toxins have been found to deliver active toxins to host cells (Horstman and Kuehn 2002). The recent findings on some specific vesicle-associated toxins are discussed below.

LT. This toxin is secreted by the enterotoxigenic *E. coli* (ETEC) cells in association with the vesicles (Horstman and Kuehn 2002). LT has been found in association with the LPS on the outer surface of the extracellular OMVs (Horstman and Kuehn 2002; Kuehn and Kesty 2005). The major fraction of LT is secreted in association with the OMVs in two ways. A fraction of LTs formed in the periplasm is capsized by the budding OMVs and retained in their lumen. Another fraction of LTs may be secreted in the extracellular medium by the type II general secretory pathway (Horstman and Kuehn 2002) as free LTs, some of which may then get attached to the LPS projecting outward from the surface of the released OMVs and thus get attached to the outer surface of the OMVs. The LTs get attached to the kdo (3-deoxy-D-manno-octulosonic acid) of the LPS where the kdo is nonphosphorylated. It was proposed that phosphorylation of kdo prevents its binding with LT (Horstman et al. 2004).

The ETEC vesicles bind via LT, enter and deliver fully active toxin into the epithelial and Y1 adrenal cells (Horstman and Kuehn 2000; Kesty et al. 2004). Several investigators have found that LT can exist both as tightly bound to LPS on the vesicles' surface and also in their lumen as a soluble cargo (Horstman et al. 2004; Horstman and Kuehn 2000, 2002; Mudrak et al. 2009).

Shiga Toxin. For different strains of Shiga toxin producing *E. coli* (STEC) the toxin was shown to be released and packaged in the lumen of the corresponding OMVs under both aerobic and anaerobic growth conditions (Kolling and Matthews 1999; Yokoyama et al. 2000). Initially the Shiga toxins (1 and/or 2) produced by the strain O157:H7 were shown to be protease resistant (Kolling and Matthews 1999), which is likely because of their presence in the lumen of OMVs. On the other hand, Shiga toxin 2 was found to bind to LPS and be neutralized (Gamage et al. 2004). This suggests that at least a fraction of the shiga toxins released may be found as bound to LPS on the surface of OMVs in a way to make it protease resistant at the same time.

CT. A recent study showed that a fraction of CT released by the toxigenic *V. cholerae* O395 was associated with the OMVs (Figs. 4.1 a, b, and c) originating from the same bacteria. Immunoblotting of purified OMVs with polyclonal anti-CT antibody and GM1-ganglioside-dependent ELISA suggested that CT was associated with OMVs (Chatterjee and Chaudhuri 2011). The authors further demonstrated by the CHO cell assay that the OMV–CT complex was physiologically active, was internalized via binding with the CT receptor (Figs. 4.2 a, b, c, and d) on the surface of the intestinal epithelial cell, and produced extra amounts of cAMP inside the epithelial cells. The OMVs thus acted as another vehicle for transport and delivery of CT into the epithelial cells. Presumably the CTs are secreted by the type II general secretory pathway into the extracellular medium and a fraction of them bind to the surface of the OMVs present there. Whether the

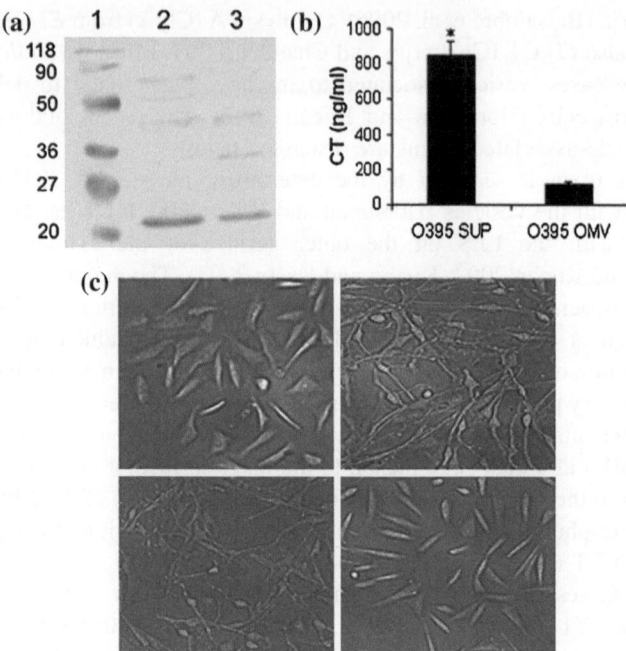

Fig. 4.1 CT association of OMV. (**a**) Immunoblot analysis of CT in *V. cholerae* O395 culture supernatant (lane 2) and OMV (lane 3) using anti-CT antibodies. Lane 1 shows molecular weight markers (kDa). (**b**) GM1-ELISA. CT was assayed in culture supernatant and intact O395 OMVs (100 μg). The data represent the average of three independent OMV preparations, each done in duplicate. The error bars indicate the standard deviation, P < 0.05. (**c**) CHO cell assay. Confluent CHO cells incubated for 12 h. Untreated conditions **a** and with 100 μg/ml purified CT as positive control **b** 100 μg/ml of O395 OMV **c** 100 μg/ml of O395 OMV preincubated with GM1 **d** and visualized under phase contrast microscope (taken from Chatterjee and Chaudhuri 2011)

CT binds to the kdo or other part of LPS of the OMVs remains to be determined, particularly because kdo in *V. cholerae* LPS is phosphorylated.

VacA. The vacuolating cytotoxin, VacA, is released by *H. pylori* through OMVs either as a full length or a natural carboxy-terminal truncated (from a VacA- phenotype strain) form. The VacA-related toxicity of OMVs from *H. pylori* has been studied at length and the OMV associated VacA was shown to cause vacuolization in HEp-2 cells (Keenan et al. 2000). However, the relative toxicity of OMV-associated versus free VacA has been debated (Ricci et al. 2005). The VacA containing OMVs were observed in the biopsied specimens of human gastric epithelium and these vesicles were very similar, both in appearance and composition, to those produced by *H. pylori* in vitro (Kadurugamuwa and Beveridge 1995). It is of interest to note that the truncated form of VacA contained the active domain but not the signal for the type I secretion process. Accordingly it was initially believed that the truncated form of VacA was a nonsecreted version of the toxin. The observation that this truncated form of VacA could be secreted

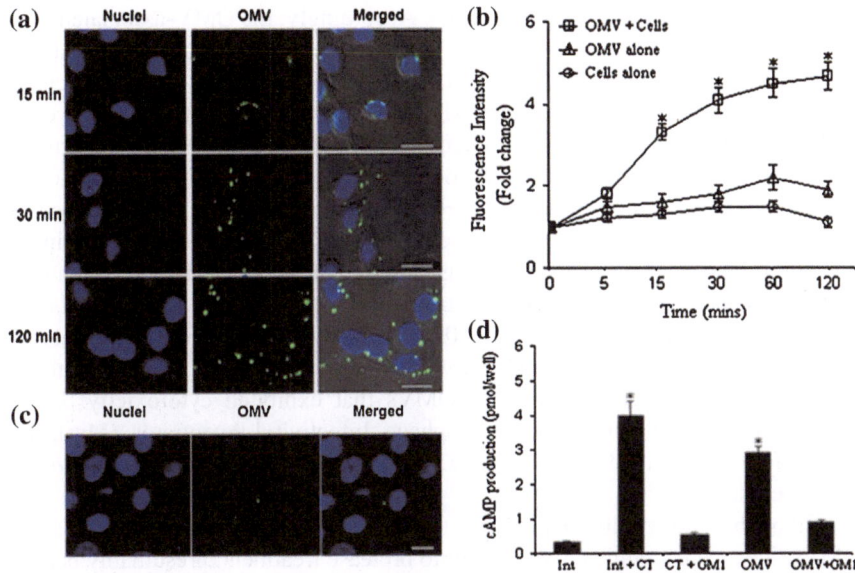

Fig. 4.2 Internalization of OMV. CT complex into epithelial cells and production of cAMP.
a Int 407 cells (approx. 4×10^4 cells) were incubated with FITC-labeled *V. cholerae* O395
OMVs (green, 20 µg/ml of protein) for 15, 30, and 120 min at 37° C. After incubation cells were
washed, fixed in 2 % paraformaldehyde. DAPI was used to stain the nuclei (*blue*) and localization
of OMVs (*green*) within Int407 cells was examined by confocal microscopy (scale
bar = 20 µm). **b** FITC-labeled vesicles were incubated with Int407 cells at 37° C and
fluorescence was measured over time as an estimation of OMV fusion. Data are presented as
mean fluorescence intensity. The error bars indicate the standard deviation, $P < 0.05$. **c** FITC-
labeled vesicles (*green*) pretreated with GM1 and incubated with Int407 cells for 2 h. At 37° C
and visualized under confocal microscopy. DAPI was used to stain the nuclei (blue). Scale
bar = 10 µm. **d** cAMP assay. Concentration of cAMP (pmol/well) produced by Int407 cells
incubated for 4 h with CT, CT preincubated with GM1, O395 vesicles, or O395 vesicles
preincubated with GM1. The error bars indicate the standard deviation, $P < 0.05$ (taken from
Chatterjee and Chaudhuri 2011)

via the OMVs and produce the cytochrome C-independent apoptosis in gastric
epithelial cells appeared surprising. It was proposed that the OMVs mediated the
toxicity of the truncated form by enabling its presentation to the host cells (Ayala
et al. 2006).

Cytolysin A (ClyA). ClyA is another toxin secreted by *E. coli* and some other
bacterial cells. It is normally found in the periplasm of the bacterial cell in an
oxidized state. With disulfide bond formation between cysteines, the ClyA
becomes unable to oligomerize and remains inactive. On the other hand, when the
ClyA is secreted in the OMVs, which are in a different redox state than the
periplasm, the reducing environment within the lumen of the OMVs, and possibly
because of their higher concentration within the OMVs, the ClyA forms an active
oligomeric pore structure (which is clearly visible by electron microscopy) on the
surface of the OMVs. Thus the OMVs reveal a unique property in maintaining the

ClyA in the active form (Wai et al. 2003). Accordingly, the OMV-associated ClyA exhibits greater toxicity than the same concentration of ClyA in the periplasm.

Other toxins. *Neisseria meningitidis* releases OMVs which are in association with a protein, NarE, that exhibits ADP-ribosyl transferase and NAD-glycohydrolase activities associated with toxicity (Masignani et al. 2003). The extraintestinal pathogenic *E. coli* (ExPEC) cells produce vesicles containing (1) α-hemolysin (hemolytic, causing detachment of cells from the monolayer), which becomes surface bound, and (2) cytolethal distending toxin (CDT), which has a lipid binding domain to bind the OM (Table 2.1; Balsalobre et al. 2006; Berlanda Scorza et al. 2008; Ellis and Kuehn 2010). Several other bacteria, including *Actinobacillus actinomycetemcomitans* (Demuth et al. 2003; Kato et al. 2002; Nowotny et al. 1982), *Xenorhabdus nematophilus*, and *Photorhabdus luminescens* (Khandelwal and Banerjee-Bhatnagar 2003) produced OMVs that exhibited cytotoxicity. Unlike VacA, the extracellular toxin, Macrophage Infectivity Potentiator (Mip), was always found associated with only the OMVs released by *Legionella pneumophila* (Galka et al. 2008).

Although many of the toxins were found associated with the outer surface of OMVs, they were significantly resistant to protease treatment, presumably because they were so tightly bound to the surface that they were not accessible to the enzyme. In fact, Leukotoxin, which kills human polymorphonuclear leukocytes and monocytes, was found tightly bound to the surface of OMVs derived from *Actinobacillus actinomycetemcomitans* (Kato et al. 2002; Ohta et al. 1993). The OMV-associated leukotoxin is more toxic than the one associated with the outer membrane (Kato et al. 2002). Furthermore, the leukotoxin has been shown to bind nucleic acid on the surface of the outer membrane (OM; Demuth et al. 2003; Ohta et al. 1991). The cytotoxins found to be tightly associated with the surface of OMVs derived from *Xenorhabdus nematophilus* (the enteric insect pathogens) could not be separated from the OMVs except by treatment with sodium dodecyl sulphate, the detergent that also destroyed its activity (Khandelwal and Banerjee-Bhatnagar 2003). These vesicles harbor chitinase activity and are more potent than the purified proteins of the (OM; Khandelwal and Banerjee-Bhatnagar 2003). The alpha hemolysin (HlyA) was also found tightly bound to the surface of OMVs derived from the ExPEC cells (Balsalobre et al. 2006; Horstman and Kuehn 2000). The mechanism that resulted in such tight binding of the toxins to the surface of OMVs remains a matter of conjecture at this moment.

4.3 Non-Toxin Virulence Factors

OMVs derived from different bacteria have been found associated with different nontoxin virulence factors, for example, adhesins, different proteases, signalling molecules, quorum sensing molecules, CiF proteins, Lewis antigens, UspA1 and UspA2, and so on. Obviously these different factors originate from different bacteria and for different purposes (see Table 2.1). Adhesins form a component of the outer membrane of OMVs and allow the OMVs to adhere to and interact with the host cells. Active proteases associated with OMVs help in degrading the host cells. OMVs derived from *Treponema denticola* contain active proteases that confer host cell damaging characteristics similar to those of the parent bacterium (Chi et al. 2003; Rosen et al. 1995; Schlichting et al. 1993). Similarly, the endotoxic filtrate of the culture supernatant of *Neisseria meningitidis* contained procoagulant and fibrinolytic factors for human monocytes (Schlichting et al. 1993). The Lewis antigens bound to the LPS of OMVs derived from *H. pylori* were supposed to contribute to the chronic stimulation of the host immune system (Hynes et al. 2005). In addition, these LPS-bound antigens can absorb anti-Lewis-antigen autoantibodies from the serum and may play a role in the putative auto-immune aspects of *H. pylori* pathogenesis. The UspA1 and UspA2 molecules associated with the OMVs derived from *M. catarrhalis* bind to C3 in the complement cascade (Tan et al. 2007) thereby inhibiting the bactericidal effects of the complement in normal human sera. These OMVs thus act simultaneously in favor of some nontypeable strains of *H. influenzae* that are also the targets of these complements. Thus by binding and depleting the complement in their immediate environment, the OMVs from one species can favor the pathogenicity of another in the same host. In addition to the above functions of the OMVs, the OMVs perhaps contain some yet undetermined factors that contribute to the bacterial entry into the host by influencing and/or increasing the bacterial cell to host cell association in a variety of tissues. Available reports say that the addition of adherent-invasive *E. coli* (AIEC) OMVs significantly increased the internalization of an OMV-underproducing mutant of AIEC (Rolhion et al. 2005). Similarly, it has been reported that the OMVs derived from *P. gingivalis* enhanced the attachment and invasion of *T. forsythia* in periodontal epithelial cells (Inagaki et al. 2006).

Both Gram-positive and Gram-negative bacteria produce quorum sensing (QS) molecules. These molecules are produced by bacteria in a population to sense cell density and to perform group behaviors accordingly. These so-called signaling molecules are synthesized by the bacteria and are then secreted by different mechanisms into the extracellular milieu. After these molecules reach a critical threshold density, they enter the cell where they bind and activate a transcriptional regulator leading to significant changes in gene expression. These molecules can also positively regulate gene expression in the parent bacteria as autoinducers (Mashburn and Whiteley 2005). Different bacteria produce different QS molecules, but Gram-negative bacteria, in general, in most cases produce the acyl homoserine lactones (HSL). All Gram-negative HSLs contain the lactone ring in common, but the length and substitution of the acyl chains may vary. Moreover, *P. aeruginosa* contain another cell-to-cell signaling molecule usually referred to as pseudomonas quinolone signal (PQS) or 2-heptyl-3-hydroxy-4-quinolone (Pesci et al. 1999), the structure of which is shown in Fig. 4.3.

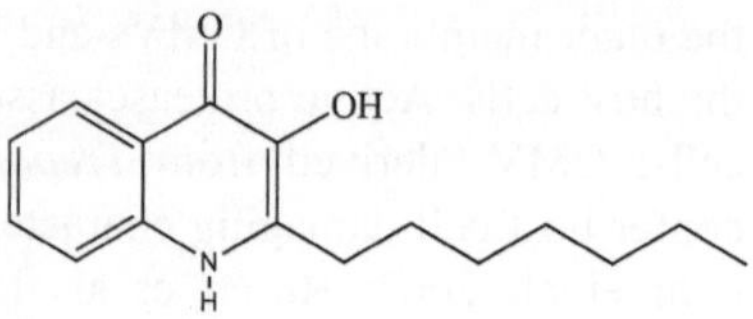

Fig. 4.3 Chemical structure
of PQS (Pseudomonas
Quinolone Signal) molecule

4.4 Enrichment of Virulence Factors in OMVs

In general, any particular constituent of the OMVs may be enriched vis-à-vis
their abundance in the parent bacterial cell, the exact reason for or the mechanism of which is not immediately clear. The protein profiles in the OM and the
OMVs may be largely similar but different in specific cases. Quantitative analysis of the contents of the bacterial OM and OMVs showed both enrichment and
depletion of the OM and periplasmic cargo (McBroom and Kuehn 2005). It has
been found that a periplasmic protein having affinity to the inner leaflet of the
outer membrane is likely to be enriched in vesicles or OMVs when compared
with the soluble periplasmic fraction of the whole cell (Kuehn and Kesty 2005).
On the other hand, a comparison of the concentration of the particular protein in
the whole cell may show that it is not enriched in the OMVs. In particular, the
virulence factors, LT of ETEC and aminopeptidase of *P. aeruginosa*, were found
enriched in OMVs. Both these factors are known to increase the association of
OMVs with cultured epithelial cells (Bernadac et al. 1998; Vasilyeva et al.
2008). For some strains of *A. actinomycetemcomitans*, the virulence factor,
leukotoxin, and its activities were found four- to fivefold enriched in the OMVs
as compared with the corresponding OM preparations (Kato et al. 2002). The
virulence factor, HlyA, was found enriched in a fraction of OMVs derived from
ExPEC cells (Balsalobre et al. 2006). On the other hand, the virulence factor,
CagA, was found to be significantly absent from the VacA-containing OMVs
derived from Cag pathogenicity island-positive strains of *H. pylori* (Ismail et al.
2003; Keenan et al. 2000). When investigated by quantitative proteomic techniques, less than half of the most abundant OM proteins of *Xanthomonas
campestris* could be detected in the corresponding OMV fraction (Sidhu et al.
2008). More detailed analysis revealed that nearly half of the proteins associated
with the OMV fraction are in some way involved in virulence, and most of the
21 proteins enriched in the OMV fraction are virulence associated. Furthermore,
the protein composition of the OMV fraction was found partially dependent on
the growth medium, indicating that the bacteria are significantly mutable by their
surroundings in the growth media with respect to the virulence factors, including
proteins and toxins, of the corresponding OMVs.

4.5 DNA

Many Gram-negative bacteria have been shown to package DNA in their OMVs (Kolling and Matthews 1999). These bacteria include *Escherichia*, *Hemophilus*, *Neisseria*, and *Pseudomonas* species (Kahn et al. 1983; Kahn et al. 1982; Dorward and Garon 1989; Dorward et al. 1989; Kadurugamuwa and Beveridge 1995; Kolling and Matthews 1999; Renelli et al. 2004; Yaron et al. 2000). The mode of DNA packaging into OMVs is largely unknown, may be a complex one, and may differ from the parent bacteria. It has also been suggested that the plasmids and chromosomal and even phage DNA can be packaged into the OMVs. DNA packaged into OMVs is protected from the extracellular DNase and is likely to enhance the DNA delivery into and uptake by a target or recipient cell. Also DNA-containing OMVs are likely to overcome the genetic barriers of noncompetent bacteria and may induce their genetic transformation (Dubnau 1999).

Experiments showed that the *P. aeruginosa* OMVs contained DNA in their lumen and that treatment with extracellular DNase could not degrade the luminal DNA (Kadurugamuwa and Beveridge 1995; Renelli et al. 2004). Kolling and Matthews (1999) demonstrated that OMVs derived from *E. coli* strain 0157:H7 were proficient in mediating transfer of DNA from one strain to another, enabling efficient transformation of genes. It was reported that the plasmid-containing OMVs could not transform the DNA into the nonplasmid-containing cells. However, OMVs from other bacteria were reported to transform DNA into target cells successfully (Renelli et al. 2004). OMVs from *N. gonorrheae* contained chromosomal DNA and plasmids of sizes 4.2 and 7.1 kb and one of them carried the penicillin-resistance gene. These OMVs also contained small amounts of RNA (Dorward et al. 1989). It was shown that penicillin-sensitive *gonococci* were transformed on incubation with OMVs containing penicillin-resistant DNA, implying thereby that OMV-mediated plasmid transfer occurred among *gonococci*. It was further shown that OMVs derived from an *E. coli* strain and containing the green fluorescent protein gene on a plasmid could transform another *E. coli* strain previously lacking the gene (Yaron et al. 2000). Renelli et al. (2004) isolated OMVs containing plasmid pAK 1900 from *P. aeruginosa* PAO1 by stringent methods to ensure no cellular contamination from the parent cell. To their surprise they found that these OMVs failed to transform neither the wild-type (lacking the particular plasmid) *P. aeruginosa* nor the *E. coli* DH5α and speculated different reasons for this failure. Their results differed from those of *N. gonorrheae* and *E. coli* 0157:H7, where the OMVs successfully transformed wild-type strains of their own species.

How DNA can be packaged into OMVs is still a matter for conjecture. For chromosomal DNA, one possibility is that DNA is first fragmented and released extracellularly by the bacteria undergoing lysis. The DNA fragments could then be picked up by the OMVs. It was shown that OMVs could pick up naked plasmid DNA and protect the DNA from subsequent treatment with DNase (Renelli et al. 2004). The other possibility is that the chromosomal DNA fragments or the plasmid DNA could be packaged from the periplasm directly into the OMVs at the stage of their budding off of the bacterial outer membrane. Both possibilities appear to be logical and are not mutually exclusive. Renelli et al. (2004) thus observed that their data favored both the possibilities of picking up DNA by the OMVs. In the context of the above discussion it may be noted that OMVs derived from *E. coli* O157:H7 were shown to package DNA from various sources, for example, chromosome, plasmids, and phages (Yaron et al. 2000). One 3.3-kb plasmid was found packaged in OMVs derived from the *E. coli* O157:H7 strain and was found to contain genes required for plasmid replication, mobilization, and partitioning but did not contain genes necessary for biosynthesis of the pilus necessary for efficient conjugation of mobilizable plasmids. The authors (Yaron et al. 2000) proposed that packaging this plasmid into OMVs resulted in its transfer from donor to recipient cells, thereby eliminating two otherwise essential requirements: (1) the need for pilus biosynthesis genes and (2) the need for close proximity of donor and recipient cells normally required for plasmid conjugal transfer. It was thus demonstrated that the OMVs provided another distinctly separate method of gene transfer (separate from transformation, conjugation, and transduction).

Are the OMVs naturally competent to uptake DNA? This aspect has also been investigated particularly using the naturally competent cells of *Hemophilus parainfluenza*. These bacteria are capable of binding and internalizing double-stranded DNA having unique sequences. A higher level of or many more OMVs were found associated with the outer surface of these bacteria than their corresponding noncompetent ones. On addition of linear DNAs to these competent cells, the corresponding OMVs disappeared. It was interpreted that the DNAs interacted with the OMVs leading to the internalization of the DNA.OMV complex (Kahn et al. 1982). Similarly *Hemophilus influenzae* cells in a competent state were shown to express numerous OMVs on the surface. But when these cells returned to normal growth or a noncompetent state, the OMVs were liberated and the bacteria underwent a significant loss of DNA binding capacity, whereas the liberated OMVs retained the DNA binding ability (Deich and Hoyer 1982). The OMVs were thus suggested to play an important role in natural competence and transformation at least in the *Hemophilus spp.*

4.6 Antibiotics

When some of the OMV-producing bacteria are treated with antibiotics, there is the chance of some antibiotics getting packaged into the corresponding OMVs. Not many antibiotics have been studied in this respect. The role of antibiotics in the formation of OMVs has already been discussed in Sect. 3.6. The role of antibiotic-loaded OMVs is therefore discussed briefly here. Of all the antibiotics, gentamicin has been most studied. When the OMV-producing *P. aeruginosa* were treated with gentamicin, some amount of the antibiotic got packaged into the corresponding OMVs thereby increasing the "predatory"activities of these OMVs (Kadurugamuwa and Beveridge 1995). These authors thus proposed that the antibiotics-loaded OMVs could be used for therapeutic purposes particularly when the naked antibiotics failed to enter the cell and kill the intracellular microbes (Kadurugamuwa and Beveridge 1998). Not many experimental trials have been done in this respect. Also, loading of OMVs with different antibiotics and their effects on the target cells invitro and also in-vivo deserve further investigation at least from the therapeutic points of view.

4.7 Engineered Recombinant OMVs

Recent studies have shown that the chemical composition including the proteomic profile of OMVs derived from a Gram-negative bacterial cell can be changed suitably to make them functionally more efficient. This can be done by incorporation of several heterologous proteins and antigens into the OMVs. Normally, different periplasmic proteins or other chemicals or different outer membrane proteins are sorted into the lumen or bounding membrane of the OMVs, respectively. Some proteins and lipids of the bacterial outer membrane or the periplasm are always found in the OMVs whereas others are always excluded. The actual mechanism of this sorting has still remained unknown. Kesty and Kuehn (2004) hypothesized that expressed heterologous outer membrane and periplasmic proteins should be packaged in vesicles and also that the vesicle properties could be altered by the expression of proteins into the periplasm and outer membrane of bacteria. It was shown that the green fluorescent protein (GFP) transported to the periplasm and packaged in vesicles could be used as a luminal vesicle marker, whereas vesicle incorporation of an outer membrane adhesin/invasin, Ail from *Yersinia enterocolitica*, could alter the adhesion and internalization properties of the vesicles vis-à-vis vesicles without Ail. When the vesicles and host cells were co-incubated the vesicles containing Ail were internalized by the eukaryotic cells, unlike those without Ail (Kesty and Kuehn 2004). The authors (Kesty and Kuehn 2004) thus demonstrated, for the first time, the incorporation of heterologously expressed outer membrane and periplasmic proteins into bacterial vesicles.

Incorporation of heterologous proteins into OMVs was reported or reviewed by several investigators (Chen et al. 2010; Collins 2011; Gorringe et al. 2009; Kim et al. 2008; Roy et al. 2010). OMVs were prepared from *Salmonella spp.* carrying *Leishmania* antigens fused to the C-terminal domain of an *E. coli* autotransporter that simultaneously integrated into the outer membrane (Schroeder and Aebischer 2009).These recombinant vesicles boosted vaccine immune responses very significantly when injected subcutaneously into mice. Similarly, heterologous proteins from other Gram-negative or even some Gram-positive bacteria could be incorporated into the OMVs by fusion with periplasmic or outer membrane proteins (Ashraf et al. 2011; Collins 2011).

The cytotoxin, ClyA, found in pathogenic and nonpathogenic *E.coli* strains and in *Salmonella enterica* serovars Typhi and Paratyphi A provides a very suitable and efficient method of producing engineered recombinant OMVs. The ClyA is a pore-forming cytotoxin (molecular weight 34 kDa) expressed by *E. coli* and some other enterobacteria. It forms stable pores in target membranes by assembling into ring- shaped toxin oligomers. Because of this pore-forming activity, ClyA lyses erythrocytes from several mammalian species. Similarly, ClyA has a lytic effect on other mammalian cells, but it has little or no effect on the bacterial cell membranes from which it is produced. Normally, ClyA is present in monomeric form in the periplasm, undergoes an oligomerization process, and forms the active pore assemblies in the OMVs and this change into the active form is dependent on the altered redox status. The oligomeric pore assemblies in the OMVs and the cyto-toxic activity of the complex towards mammalian cells are considerably higher than that of ClyA protein purified from the bacterial periplasm. Activation of a gene *clyA* results in expression of ClyA protein. Intact *E. coli* cells expressing ClyA exhibit a cytotoxic effect upon direct contact with mammalian cells and ClyA's activity appears to be contact dependent. ClyA is a long, rod-shaped molecule, with approximate dimensions of $100 \times 30 \times 20$ Å. The shaft of the rod is formed by a bundle of four helices. Crystallographic studies indicate that the water-soluble form of ClyA is a bundle of four major α-helices (Fig. 4.4), with a small surface exposed hydrophobic ß-hairpin at the head-end of the structure, and the N- and C-termini at the tail end (Wallace et al. 2000). The lipid-associated ClyA forms an oligomeric pore complex composed of either 8 or 13 subunits (Tzokov et al. 2006; Eifler et al. 2006). ClyA is exported from *E. coli* in OMVs and retains a cytolytically active, oligomeric conformation in the vesicles (Wai et al. 2003). But the exact role ClyA plays in vesicle-mediated interactions with mammalian cells is not known. It was suggested that the assembly of ClyA oligomers in OMVs should be considered as a process, which exports ClyA to the external environment. It is feasible that the OMVs containing ClyA were fusing with the target cells and that the cytotoxin thereby was also more efficiently delivered to the target cell membranes.

Engineered recombinant OMVs with strikingly enhanced functionality were obtained by fusing several heterologous proteins to the vesicle-associated toxin, ClyA, of *E. coli* (Kim et al. 2008). ClyA could be used to co-localize fully functional heterologous proteins directly in bacterial OMVs. Fusion of ClyA to the enzymes ß-lactamase and organophosphorus hydrolase resulted in recombinant synthetic OMVs that were capable of hydrolyzing ß-lactum antibiotics and paraoxon, respectively. OMVs displaying green fluorescent protein fused to the C-terminus of ClyA were highly fluorescent and accordingly could be easily tracked during vesicle interaction with human epithelial cells (Kim et al. 2008). Using the green fluorescent protein as the model subunit antigen, genetic fusion of GFP with the bacterial hemolysin ClyA resulted in a chimeric protein that elicited strong anti-GFP antibody titers in immunized mice, whereas immunization with GFP alone did not elicit such titers (Chen et al. 2010). Although ClyA is efficiently translocated across the inner or cytoplasmic membrane to the periplasm and the

Fig. 4.4 Structure of
monomeric ClyA as derived
by X-Ray crystallography
(from RSCB-PDB Database
as viewed by Jmol 3D
viewer)

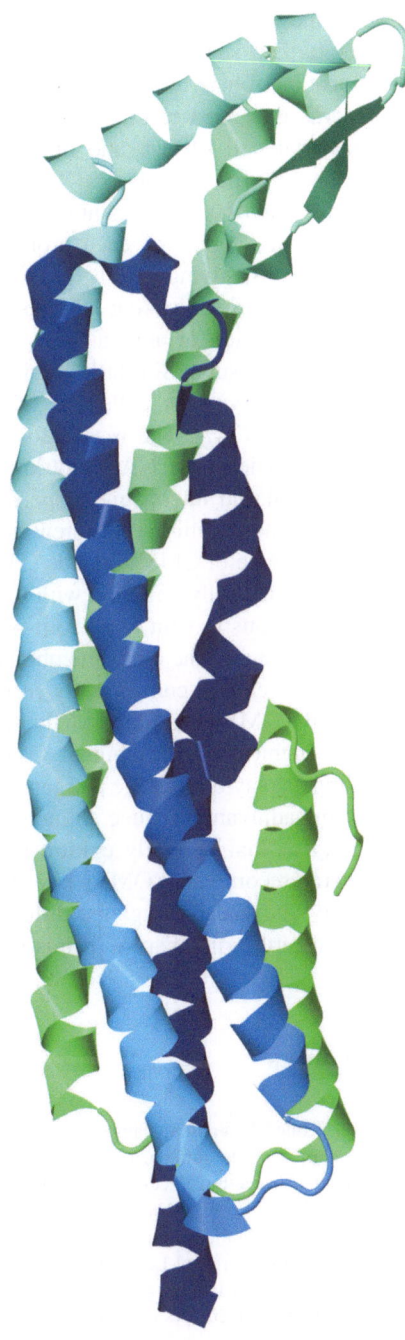

OMVs, the actual mechanism of translocation is not known (Kim et al. 2008). It was further shown that genetic fusion with ClyA resulted in the efficient translocation of heterologous proteins to OMVs (Kim et al. 2008). It was observed, in general, that most recombinant polypeptide fusions co-localized with ClyA to the bacterial cell surface and into OMVs. Direct fusion of β-lactamase (Bla), organophosphorus hydrolase (OPH), green fluorescent protein, and antidigoxin single-chain antibody fragment (ScFv.Dig) to the C-terminus of ClyA resulted in the functional display of each protein on the surface of *E. coli* cells and their derived OMVs, giving rise to recombinant OMVs with significantly expanded, non-native functionality (e.g., fluorescence, antigen binding, etc.). It is interesting to note that fusion of any of these proteins to the N-terminus of ClyA yielded unpredictable results. Similarly, fusion of the enzymes Bla and OPH resulted in little to no activity on cells and OMVs when each enzyme was fused to the N-terminus of ClyA. The reason for this difference between the C-terminal and N-terminal fusions to ClyA is not known at present.

The useful and interesting properties of ClyA that make it a very convenient vehicle for translocation of different heterologous proteins, antigens , and the like, to the membrane of OMVs include (1) OMVs derived from *E. coli* and some other bacteria are naturally enriched with ClyA, (2) ClyA alone or in combination with any heterologous protein fused to it is efficiently transported across the plasma or inner membrane to the outer membrane and the OMVs, and (3) different proteins and antigens can be conveniently fused to the C-terminus of ClyA for transportation to the OMVs in a functionally active state. Thus by using ClyA as the fusion partner, engineered recombinant OMVs can be prepared for various purposes including display of multiple antigens on the vesicle surface, can function both as a career and adjuvant, produce an increase in the immunogenic property of proteins and antigens that normally exhibit low immunogenic properties and so on. The engineered recombinant OMVs have thus a great role to play in the production of immunity and vaccines against different human and animal pathogens, which is discussed further in Chap. 9 of this monograph.

References

Ashraf S, Kong W, Wang S, Yang J, Curtiss R 3rd (2011) Protective cellular responses elicited by vaccination with influenza nucleoprotein delivered by a live recombinant attenuated *Salmonella* vaccine. Vaccine 29:3990–4002

Ayala G, Torres L, Espinosa M, Fierros-Zarate G, Maldonado V, Melendez-Zajgla J (2006) External membrane vesicles from *Helicobacter pylori* induce apoptosis in gastric epithelial cells. FEMS Microbiol Lett 260:178–185

Balsalobre C, Silvan JM, Berglund S, Mizunoe Y, Uhlin BE, Wai SN (2006) Release of the type I secreted alpha-haemolysin via outer membrane vesicles from *Escherichia coli*. Mol Microbiol 59:99–112

Berlanda Scorza F, Doro F, Rodriguez-Ortega MJ, Stella M, Liberatori S et al (2008) Proteomics characterization of outer membrane vesicles from the extraintestinal pathogenic *Escherichia coli* DeltatolR IHE3034 mutant. Mol Cell Proteomics 7:473–485

Bernadac A, Gavioli M, Lazzaroni JC, Raina S, Lloubes R (1998) *Escherichia coli* tol-pal mutants form outer membrane vesicles. J Bacteriol 180:4872–4878

Beveridge TJ, Makin SA, Kadurugamuwa JL, Li Z (1997) Interactions between biofilms and the environment. FEMS Microbiol Rev 20:291–303

Chatterjee D, Chaudhuri K (2011) Association of cholera toxin with *Vibrio cholerae* outer membrane vesicles which are internalized by human intestinal epithelial cells. FEBS Lett 585:1357–1362

Chen DJ, Osterrieder N, Metzger SM, Buckles E, Doody AM, DeLisa MP, Putnam D (2010) Delivery of foreign antigens by engineered outer membrane vesicle vaccines. Proc Natl Acad Sci USA 107:3099–3104

Chi B, Qi M, Kuramitsu HK (2003) Role of dentilisin in *Treponema denticola* epithelial cell layer penetration. Res Microbiol 154:637–643

Collins BS (2011) Gram-negative outer membrane vesicles in vaccine development. Discov Med 12:7–15

Deich RA, Hoyer LC (1982) Generation and release of DNA-binding vesicles by *Hemophilus influenzae* during induction and loss of competence. J Bacteriol 152:855–864

Demuth DR, James D, Kowashi Y, Kato S (2003) Interaction of *Actinobacillus actinomycetemcomitans* outer membrane vesicles with HL60 cells does not require leukotoxin. Cell Microbiol 5:111–121

Dorward DW, Garon CF (1989) DNA-binding proteins in cells and membrane blebs of *Neisseria gonorrhoeae*. J Bacteriol 171:4196–4201

Dorward DW, Garon CF, Judd RC (1989) Export and intercellular transfer of DNA via membrane blebs of *Neisseria gonorrhoeae*. J Bacteriol 171:2499–2505

Dubnau D (1999) DNA uptake in bacteria. Annu Rev Microbiol 53:217–244

Eifler N, Vetsch M, Gregorini M, Ringler P, Chami M et al (2006) Cytotoxin ClyA from *Escherichia coli* assembles to a 13-meric pore independent of its redox-state. EMBO J 25:2652–2661

Ellis TN, Kuehn MJ (2010) Virulence and immunomodulatory roles of bacterial outer membrane vesicles. Microbiol Mol Biol Rev 74:81–94

Galka F, Wai SN, Kusch H, Engelmann S, Hecker M et al (2008) Proteomic characterization of the whole secretome of *Legionella pneumophila* and functional analysis of outer membrane vesicles. Infect Immun 76:1825–1836

Gamage SD, McGannon CM, Weiss AA (2004) *Escherichia coli* serogroup O107/O117 lipopolysaccharide binds and neutralizes Shiga toxin 2. J Bacteriol 186:5506–5512

Gorringe AR, Taylor S, Brookes C, Matheson M, Finney M et al (2009) Phase I safety and immunogenicity study of a candidate meningococcal disease vaccine based on *Neisseria lactamica* outer membrane vesicles. Clin Vaccine Immunol 16:1113–1120

Hoekstra D, van der Laan JW, de Leij L, Witholt B (1976) Release of outer membrane fragments from normally growing *Escherichia coli*. Biochim Biophys Acta 455:889–899

Horstman AL, Kuehn MJ (2000) Enterotoxigenic *Escherichia coli* secretes active heat-labile enterotoxin via outer membrane vesicles. J Biol Chem 275:12489–12496

Horstman AL, Kuehn MJ (2002) Bacterial surface association of heat-labile enterotoxin through lipopolysaccharide after secretion via the general secretory pathway. J Biol Chem 277:32538–32545

Horstman AL, Bauman SJ, Kuehn MJ (2004) Lipopolysaccharide 3-deoxy-D-manno-octulosonic acid (Kdo) core determines bacterial association of secreted toxins. J Biol Chem 279:8070–8075

Hynes SO, Keenan JI, Ferris JA, Annuk H, Moran AP (2005) Lewis epitopes on outer membrane vesicles of relevance to *Helicobacter pylori* pathogenesis. Helicobacter 10:146–156

Inagaki S, Onishi S, Kuramitsu HK, Sharma A (2006) *Porphyromonas gingivalis* vesicles enhance attachment, and the leucine-rich repeat BspA protein is required for invasion of epithelial cells by "*Tannerella forsythia*". Infect Immun 74:5023–5028

Ismail S, Hampton MB, Keenan JI (2003) *Helicobacter pylori* outer membrane vesicles modulate proliferation and interleukin-8 production by gastric epithelial cells. Infect Immun 71: 5670–5675

Kadurugamuwa JL, Beveridge TJ (1995) Virulence factors are released from *Pseudomonas aeruginosa* in association with membrane vesicles during normal growth and exposure to gentamicin: a novel mechanism of enzyme secretion. J Bacteriol 177:3998–4008

Kadurugamuwa JL, Beveridge TJ (1996) Bacteriolytic effect of membrane vesicles from *Pseudomonas aeruginosa* on other bacteria including pathogens: conceptually new antibiotics. J Bacteriol 178:2767–2774

Kadurugamuwa JL, Beveridge TJ (1998) Delivery of the non-membrane-permeative antibiotic gentamicin into mammalian cells by using *Shigella flexneri* membrane vesicles. Antimicrob Agent Chemother 42:1476–1483

Kahn ME, Maul G, Goodgal SH (1982) Possible mechanism for donor DNA binding and transport in *Hemophilus*. Proc Natl Acad Sci USA 79:6370–6374

Kahn ME, Barany F, Smith HO (1983) Transformasomes: specialized membranous structures that protect DNA during *Hemophilus* transformation. Proc Natl Acad Sci USA 80:6927–6931

Kato S, Kowashi Y, Demuth DR (2002) Outer membrane-like vesicles secreted by *Actinobacillus actinomycetemcomitans* are enriched in leukotoxin. Microb Pathog 32:1–13

Keenan J, Day T, Neal S, Cook B, Perez–Perez G, Allardyce R, Bagshaw P (2000) A role for the bacterial outer membrane in the pathogenesis of *Helicobacter pylori* infection. FEMS Microbiol Lett 182:259–264

Kesty NC, Kuehn MJ (2004) Incorporation of heterologous outer membrane and periplasmic proteins into *Escherichia coli* outer membrane vesicles. J Biol Chem 279:2069–2076

Kesty NC, Mason KM, Reedy M, Miller SE, Kuehn MJ (2004) Enterotoxigenic *Escherichia coli* vesicles target toxin delivery into mammalian cells. EMBO J 23:4538–4549

Khandelwal P, Banerjee-Bhatnagar N (2003) Insecticidal activity associated with the outer membrane vesicles of *Xenorhabdus nematophilus*. Appl Environ Microbiol 69:2032–2037

Kim JY, Doody AM, Chen DJ, Cremona GH, Shuler ML, Putnam D, DeLisa MP (2008) Engineered bacterial outer membrane vesicles with enhanced functionality. J Mol Biol 380:51–66

Kolling GL, Matthews KR (1999) Export of virulence genes and Shiga toxin by membrane vesicles of *Escherichia coli* O157:H7. Appl Environ Microbiol 65:1843–1848

Kuehn MJ, Kesty NC (2005) Bacterial outer membrane vesicles and the host-pathogen interaction. Genes Dev 19:2645–2655

Li Z, Clarke AJ, Beveridge TJ (1996) A major autolysin of *Pseudomonas aeruginosa*: subcellular distribution, potential role in cell growth and division and secretion in surface membrane vesicles. J Bacteriol 178:2479–2488

Li Z, Clarke AJ, Beveridge TJ (1998) Gram-negative bacteria produce membrane vesicles which are capable of killing other bacteria. J Bacteriol 180:5478–5483

Mashburn LM, Whiteley M (2005) Membrane vesicles traffic signals and facilitate group activities in a prokaryote. Nature 437:422–425

Mashburn-Warren LM, Whiteley M (2008) Signal trafficking with bacterial outer membrane vesicles. In: Winans SC, Bassler BL (eds) Chemical communication among bacteria. ASM Press, Washington

Masignani V, Balducci E, Di Marcello F, Savino S, Serruto D et al (2003) NarE: a novel ADP-ribosyltransferase from *Neisseria meningitidis*. Mol Microbiol 50:1055–1067

McBroom AJ, Kuehn MJ (2005) Outer membrane vesicles. In: RC III (ed) EcoSal - *Escherichia coli* and *Salmonella* : Cellular and Molecular Biology. American Society for Microbiology Press, Washington

Mudrak B, Rodriguez DL, Kuehn MJ (2009) Residues of heat-labile enterotoxin involved in bacterial cell surface binding. J Bacteriol 191:2917–2925

Mug-Opstelten D, Witholt B (1978) Preferential release of new outer membrane fragments by exponentially growing *Escherichia coli*. Biochim Biophys Acta 508:287–295

Negrete-Abascal E, Garcia RM, Reyes ME, Godinez D, de la Garza M (2000) Membrane vesicles released by *Actinobacillus pleuropneumoniae* contain proteases and Apx toxins. FEMS Microbiol Lett 191:109–113

Nowotny A, Behling UH, Hammond B, Lai CH, Listgarten M, Pham PH, Sanavi F (1982) Release of toxic microvesicles by *Actinobacillus actinomycetemcomitans*. Infect Immun 37:151–154

Ohta H, Kato K, Kokeguchi S, Hara H, Fukui K, Murayama Y (1991) Nuclease-sensitive binding of an *Actinobacillus actinomycetemcomitans* leukotoxin to the bacterial cell surface. Infect Immun 59:4599–4605

Ohta H, Hara H, Fukui K, Kurihara H, Murayama Y, Kato K (1993) Association of *Actinobacillus actinomycetemcomitans* leukotoxin with nucleic acids on the bacterial cell surface. Infect Immun 61:4878–4884

Pesci EC, Milbank JB, Pearson JP, McKnight S, Kende AS, Greenberg EP, Iglewski BH (1999) Quinolone signaling in the cell-to-cell communication system of *Pseudomonas aeruginosa*. Proc Natl Acad Sci USA 96:11229–11234

Renelli M, Matias V, Lo RY, Beveridge TJ (2004) DNA-containing membrane vesicles of *Pseudomonas aeruginosa* PAO1 and their genetic transformation potential. Microbiology 150:2161–2169

Ricci V, Chiozzi V, Necchi V, Oldani A, Romano M, Solcia E, Ventura U (2005) Free-soluble and outer membrane vesicle-associated VacA from *Helicobacter pylori*: Two forms of release, a different activity. Biochem Biophys Res Commun 337:173–178

Rolhion N, Barnich N, Claret L, Darfeuille-Michaud A (2005) Strong decrease in invasive ability and outer membrane vesicle release in Crohn's disease-associated adherent-invasive *Escherichia coli* strain LF82 with the yfgL gene deleted. J Bacteriol 187:2286–2296

Rosen G, Naor R, Rahamim E, Yishai R, Sela MN (1995) Proteases of *Treponema denticola* outer sheath and extracellular vesicles. Infect Immun 63:3973–3979

Roy N, Barman S, Ghosh A, Pal A, Chakraborty K et al (2010) Immunogenicity and protective efficacy of *Vibrio cholerae* outer membrane vesicles in rabbit model. FEMS Immunol Med Microbiol 60:18–27

Schlichting E, Lyberg T, Solberg O, Andersen BM (1993) Endotoxin liberation from *Neisseria meningitidis* correlates to their ability to induce procoagulant and fibrinolytic factors in human monocytes. Scand J Infect Dis 25:585–594

Schroeder J, Aebischer T (2009) Recombinant outer membrane vesicles to augment antigen-specific live vaccine responses. Vaccine 27:6748–6754

Sidhu VK, Vorholter FJ, Niehaus K, Watt SA (2008) Analysis of outer membrane vesicle associated proteins isolated from the plant pathogenic bacterium *Xanthomonas campestris pv. campestris*. BMC Microbiol 8:87

Tan TT, Morgelin M, Forsgren A, Riesbeck K (2007) *Hemophilus influenzae* survival during complement-mediated attacks is promoted by Moraxella catarrhalis outer membrane vesicles. J Infect Dis 195:1661–1670

Tzokov SB, Wyborn NR, Stillman TJ, Jamieson S, Czudnochowski N et al (2006) Structure of the hemolysin E (HlyE, ClyA, and SheA) channel in its membrane-bound form. J Biol Chem 281:23042–23049

Vasilyeva NV, Tsfasman IM, Suzina NE, Stepnaya OA, Kulaev IS (2008) Secretion of bacteriolytic endopeptidase L5 of Lysobacter sp. XL1 into the medium by means of outer membrane vesicles. FEBS J 275:3827–3835

Wai SN, Lindmark B, Soderblom T, Takade A, Westermark M et al (2003) Vesicle-mediated export and assembly of pore-forming oligomers of the enterobacterial ClyA cytotoxin. Cell 115:25–35

Wallace AJ, Stillman TJ, Atkins A, Jamieson SJ, Bullough PA, Green J, Artymiuk PJ (2000) *E. coli* hemolysin E (HlyE, ClyA, SheA): X-ray crystal structure of the toxin and observation of membrane pores by electron microscopy. Cell 100:265–276

Yaron S, Kolling GL, Simon L, Matthews KR (2000) Vesicle-mediated transfer of virulence
    genes from *Escherichia coli* O157:H7 to other enteric bacteria. Appl Environ Microbiol
    66:4414–4420

Yokoyama K, Horii T, Yamashino T, Hashikawa S, Barua S et al (2000) Production of shiga
    toxin by *Escherichia coli* measured with reference to the membrane vesicle-associated toxins.
    FEMS Microbiol Lett 192:139–144

Chapter 5
Outer Membrane Vesicles and the Biofilm Formation

Abstract Plenty of outer membrane vesicles (OMVs) are often found associated with the parent bacteria forming a biofilm. The functions of these OMVs in the biofilm are discussed in this chapter. Additional genetic and other studies are required to decide the exact role of OMVs and DNA in the formation of biofilm.

Keywords OMVs · Biofilm · DNA · Genetics · LPS

Bacteria can live either as free-living planktonic organisms or in biofilms. A biofilm is a sheet of aggregated bacteria attached to a surface and is produced by a self-produced exopolymerase matrix. In this state, the bacteria are protected from a variety of extracellular agents including antibiotics and immune factors (Costerton et al. 1995; Costerton et al. 1999; Lawrence et al. 1991; Ramsey and Whiteley 2004; Mashburn-Warren 2008). The biofilms formed in vivo or particularly within a host (human, animal, etc.) have thus been an important subject of study particularly with respect to microbial pathogenesis. The biofilms produced in vitro or in vivo by Gram-negative bacteria are often found to contain plenty of OMVs. It is generally believed that the OMVs have a significant role in the formation of biofilms.

OMVs were found to mediate co-aggregation of bacteria enabling biofilm formation and colonization (Grenier and Mayrand 1987; Whitchurch et al. 2002). OMVs can provide interbacterial glue to generate a nearly impervious multicellular structure, such as in biofilms, thereby producing resistance to antibiotics and other antibacterial agents. The *P. gingivalis* bacteria express the protein, HmuY, for utilization of heme which was found associated with OMVs and to promote biofilm formation (Olczak et al. 2010; Unal et al. 2010). It was hypothesized that OMVs being associated with biofilms are an important part of the microflora (Kulp and Kuehn 2010). In a situation when nutrients were scarce and during hostile growth conditions the OMVs were reported to lyse the foreign bacteria there by supporting the growth of the original ones (Kulp and Kuehn 2010). The OMVs

S. N. Chatterjee and K. Chaudhuri, *Outer Membrane Vesicles of Bacteria*, SpringerBriefs in Microbiology, DOI: 10.1007/978-3-642-30526-9_5, © The Author(s) 2012

originating from *P. aeruginosa* contained the enzyme peptidoglycan hydrolase, which could lyse the competing bacteria and give survival advantage to the parent ones (Kulp and Kuehn 2010; MacDonald and Beveridge 2002). Similarly, several organisms including *P. aeruginosa* and *M. catarrhalis* were found to produce proteins of the family ß-lactamase, which could break down antibiotics such as amoxicillin containing the ß-lactum ring in their structures (Bomberger et al. 2009). This could be one mechanism for producing antibiotic resistance of bacteria in biofilms.

Many oral bacteria were found to promote biofilm formation and colonization (Ellis and Kuehn 2010; Grenier and Mayrand 1987; Kamaguchi et al. 2003a, b; Mayrand and Grenier 1989). *Porphyromonas gingivalis*, one of the bacteria responsible for periodontal disease, produces vesicles with the potential to induce auto-aggregation and co-aggregation with other bacteria. In vitro, *P. gingivalis* OMVs co-aggregated various oral microorganisms including *Eubacterium saburreum*, *Capnocytophaga ochracea*, *Staphylococcus aureus*, various *Streptococcus* species, and so on (Grenier and Mayrand 1987; Kamaguchi et al. 2003a, b). The proteolytic and adhesive activities of *P. gingivalis* are due to the presence of gingipains (Grenier and Mayrand 1987; Kamaguchi et al. 2003a, b; Patrick et al. 1996). The periodonto-pathogen, *Tannerella forsythia* was found to co-aggregate by the action of OMVs from *P. gingivalis*. Also, the interaction or aggregation of *Staphylococci* and *Prevotella intermedia* were mediated by OMVs from *Tannerella forsythia* (Ellis and Kuehn 2010; Inagaki et al. 2006; Kamaguchi et al. 2003a, b). Thus, vesicles could have an impact on both the establishment and the longevity of the infecting bacterial population. It was suggested that DNA bound to the surface of OMVs provided a bridging component that led to interactions among biofilm matrix, bacteria, and OMVs (Schooling and Beveridge 2006; Schooling et al. 2009).

DNA was found to be a consistent component of biofilm biomass and it was suggested that the DNA-containing OMVs might contribute to this (Renelli et al. 2004). Thus the matrix that glues the biofilm together is composed of exopolysaccharides, proteins, and DNA. It was shown for alginet-producing *P. aeruginosa* that a consistent proportion of the extracellular material was DNA (Whitchurch et al. 2002). Addition of DNase to the established biofilm resulted in the destruction of many early biofilms. It was argued that because little cell lysis was observed in these biofilms, at least a fraction of this DNA might be due to the existence of DNA associated with OMVs. The biofilm formation may thus be enhanced by the OMVs containing DNA (Renelli et al. 2004). However, the exact chemical nature of the interactions among biofilm matrix, bacteria, and OMVs with DNA acting as a bridging component remains to be worked out.

The genetics of biofilm formation and particularly the role of OMVs in its formation have not been adequately studied but are important particularly with respect to regulation of such processes in vivo and also in vitro (Chatterjee and Chaudhuri 2006; Chaudhuri and Chatterjee 2009). It was shown, regarding several strains of *V. cholerae* that the development of a three-dimensional biofilm required the presence of *vps* genes, which are responsible for the synthesis of an exo polysaccharide-based adhesive extracellular matrix (Watnick et al. 1999; Watnick and Kolter 1999; Yildiz and Schoolnik 1999).Watnick and Kolter (1999) further reported, using transposon mutagenesis, that the genes involved in biofilm formation included those encoding (1) the biosynthesis and secretion of the type IV pilus (mannose-sensitive hemagglutinin pilus MSHA), (2) the synthesis of exopolysaccharide, and (3) flagellar motility. Accordingly, they suggested that the three steps in the process of biofilm formation were: (1) type IV pilus and the flagellum accelerate attachment to the abiotic surface, (2) flagellum mediates the spread along the abiotic surface, and (3) exopolysaccharide forms the three-dimensional biofilm architecture. The biofilm formation is normally associated with the change from a normal, smooth colony morphology to the rugose one of the bacteria (Mizunoe et al. 1999). The rugose colony morphology was the result of increased synthesis of the VPS exopolysaccharide (Wai et al. 1998; Yildiz and Schoolnik 1999), and the transcriptional regulation of the *vps* genes, which are required for the synthesis of the VPS exopolysaccharide, was altered in these strains (Yildiz et al. 2001). Thus, these variants rapidly formed biofilms in LB broth that were much thicker than those formed by smooth-colony variants of. *V. cholerae*. *V. cholerae* O139 strain MO10 was also shown to produce exopolysaccharide leading to biofilm formation in response to nutrient starvation.

The influence of biofilm formation on the structure of lipopolysaccharide (LPS) or vice versa is another important field of study. In *Pseudomonas aeruginosa*, studies have indicated that changes in LPS phenotype affected adherence properties and influenced biofilm formation (Rocchetta et al. 1999). Nesper and co-workers studied several aspects, including resistance to phage K139.cm9 of and biofilm formation by the different *galU* and *galE* mutants of *V. cholerae* biotype El Tor (Nesper et al. 2001). Among the spontaneous phage K139.cm9-resistant strains, they found strains with rugose colony morphology that constitutively synthesized an exopolysaccharide and produced biofilm on abiotic surfaces. They introduced *galU* and *galE* mutations into the rugose variant P27459res105 and found that both mutations yielded smooth colony forms, suggesting that *galU* and *galE* mutants were unable to synthesize the exopolysaccharide and could not form the biofilm. The fact that *galU* and *galE* were found essential for the formation of a biofilm by the phage-resistant rugose-variant strain suggested that the synthesis of UDP-galactose via UDP-glucose was necessary for the biosynthesis of exopolysaccharide. Kierek and Watnik reported the formation of *vps*-independent biofilm of *V. cholerae* in model sea water. Although Ca^{+2} was shown to be required for the formation of *vps*-independent biofilm (Kierek and Watnick 2003), the exact mechanism remains to be worked out. It was, however, shown that (1) both MSHA and flagellum were required for the formation of *vps*-independent biofilm, (2) both the O-antigen and capsule of *V. cholerae* O139 promoted this biofilm formation, (3) spontaneous unencapsulated variants of O139 also exhibited markedly increased surface association, (4) Ca^{+2} was an integral component of the *vps*-independent extracellular biofilm matrix, and (5) the biofilm formed in true sea water exhibited O-antigen polysaccharide dependence and disintegrated upon exposure to true fresh water. LPS was thus found to play a significant role in biofilm formation. However, the genetics of the role of OMVs in biofilm formation by different organisms need to be studied in detail.

References

Bomberger JM, Maceachran DP, Coutermarsh BA, Ye S, O'Toole GA, Stanton BA (2009) Long-distance delivery of bacterial virulence factors by *Pseudomonas aeruginosa* outer membrane vesicles. PLoS Pathog 5:e1000382

Chatterjee SN, Chaudhuri K (2006) Lipopolysaccharides of *Vibrio cholerae*: III. Biological functions. Biochim Biophys Acta 1762:1–16

Chaudhuri K, Chatterjee SN (2009) Cholera toxins. Springer, Berlin

Costerton JW, Lewandowski Z, Caldwell DE, Korber DR, Lappin-Scott HM (1995) Microbial biofilms. Annu Rev Microbiol 49:711–745

Costerton JW, Stewart PS, Greenberg EP (1999) Bacterial biofilms: a common cause of persistent infections. Science 284:1318–1322

Ellis TN, Kuehn MJ (2010) Virulence and immunomodulatory roles of bacterial outer membrane vesicles. Microbiol Mol Biol Rev 74:81–94

Grenier D, Mayrand D (1987) Functional characterization of extracellular vesicles produced by *Bacteroides gingivalis*. Infect Immun 55:111–117

Inagaki S, Onishi S, Kuramitsu HK, Sharma A (2006) *Porphyromonas gingivalis* vesicles enhance attachment, and the leucine-rich repeat BspA protein is required for invasion of epithelial cells by "*Tannerella forsythia*". Infect Immun 74:5023–5028

Kamaguchi A, Nakayama K, Ichiyama S, Nakamura R, Watanabe T et al (2003a) Effect of *Porphyromonas gingivalis* vesicles on coaggregation of *Staphylococcus aureus* to oral microorganisms. Curr Microbiol 47:485–491

Kamaguchi A, Ohyama T, Sakai E, Nakamura R, Watanabe T, Baba H, Nakayama K (2003b) Adhesins encoded by the gingipain genes of *Porphyromonas gingivalis* are responsible for co-aggregation with *Prevotella intermedia*. Microbiology 149:1257–1264

Kierek K, Watnick PI (2003) The *Vibrio cholerae* O139 O-antigen polysaccharide is essential for Ca2 + -dependent biofilm development in sea water. Proc Natl Acad Sci USA 100: 14357–14362

Kulp A, Kuehn MJ (2010) Biological functions and biogenesis of secreted bacterial outer membrane vesicles. Annu Rev Microbiol 64:163–184

Lawrence JR, Korber DR, Hoyle BD, Costerton JW, Caldwell DE (1991) Optical sectioning of microbial biofilms. J Bacteriol 173:6558–6567

MacDonald KL, Beveridge TJ (2002) Bactericidal effect of gentamicin-induced membrane vesicles derived from *Pseudomonas aeruginosa* PAO1 on gram-positive bacteria. Can J Microbiol 48:810–820

Mashburn-Warren L (2008) Quinolone trafficking via outer membrane vesicles in Pseudomonas aeruginosa. Dissertation for the Doctor of Philosophy Thesis, University of Texas, Austin

Mayrand D, Grenier D (1989) Biological activities of outer membrane vesicles. Can J Microbiol 35:607–613

Mizunoe Y, Wai SN, Takade A, Yoshida SI (1999) Isolation and characterization of rugose form of *Vibrio cholerae* O139 strain MO10. Infect Immun 67:958–963

Nesper J, Lauriano CM, Klose KE, Kapfhammer D, Kraiss A, Reidl J (2001) Characterization of *Vibrio cholerae* O1 El tor galU and galE mutants: influence on lipopolysaccharide structure, colonization, and biofilm formation. Infect Immun 69:435–445

Olczak T, Wojtowicz H, Ciuraszkiewicz J, Olczak M (2010) Species specificity, surface exposure, protein expression, immunogenicity, and participation in biofilm formation of *Porphyromonas gingivalis* HmuY. BMC Microbiol 10:134

Patrick S, McKenna JP, O'Hagan S, Dermott E (1996) A comparison of the haemagglutinating and enzymic activities of Bacteroides fragilis whole cells and outer membrane vesicles. Microb Pathog 20:191–202

Ramsey MM, Whiteley M (2004) *Pseudomonas aeruginosa* attachment and biofilm development in dynamic environments. Mol Microbiol 53:1075–1087

Renelli M, Matias V, Lo RY, Beveridge TJ (2004) DNA-containing membrane vesicles of *Pseudomonas aeruginosa* PAO1 and their genetic transformation potential. Microbiology 150:2161–2169

Rocchetta HL, Burrows LL, Lam JS (1999) Genetics of O-antigen biosynthesis in *Pseudomonas aeruginosa*. Microbiol Mol Biol Rev 63:523–553

Schooling SR, Beveridge TJ (2006) Membrane vesicles: an overlooked component of the matrices of biofilms. J Bacteriol 188:5945–5957

Schooling SR, Hubley A, Beveridge TJ (2009) Interactions of DNA with biofilm-derived membrane vesicles. J Bacteriol 191:4097–4102

Unal CM, Schaar V, Riesbeck K (2010) Bacterial outer membrane vesicles in disease and preventive medicine. Semin Immunopathol 33:395–408

Wai SN, Mizunoe Y, Takade A, Kawabata SI, Yoshida SI (1998) *Vibrio cholerae* O1 strain TSI-4 produces the exopolysaccharide materials that determine colony morphology, stress resistance, and biofilm formation. Appl Environ Microbiol 64:3648–3655

Watnick PI, Kolter R (1999) Steps in the development of a *Vibrio cholerae* El Tor biofilm. Mol Microbiol 34:586–595

Watnick PI, Fullner KJ, Kolter R (1999) A role for the mannose-sensitive hemagglutinin in biofilm formation by *Vibrio cholerae* El Tor. J Bacteriol 181:3606–3609

Whitchurch CB, Tolker-Nielsen T, Ragas PC, Mattick JS (2002) Extracellular DNA required for bacterial biofilm formation. Science 295:1487

Yildiz FH, Schoolnik GK (1999) *Vibrio cholerae* O1 El Tor: identification of a gene cluster required for the rugose colony type, exopolysaccharide production, chlorine resistance, and biofilm formation. Proc Natl Acad Sci USA 96:4028–4033

Yildiz FH, Dolganov NA, Schoolnik GK (2001) VpsR, a Member of the Response Regulators of the Two-Component Regulatory Systems, Is Required for Expression of vps Biosynthesis Genes and EPS(ETr)-Associated Phenotypes in *Vibrio cholerae* O1 El Tor. J Bacteriol 183:1716–1726

Chapter 6
Outer Membrane Vesicles: Interaction with Prokaryotes and Eukaryotes

Abstract In order to favour the parent bacteria by their predatory activities, the outer membrane vesicles (OMVs) have to interact with other competing prokaryotes (Gram-negative or Gram-positive) or with the eukaryotic cells as hosts. The mechanisms of action of the OMVs including their modes of binding with and delivery of packaged materials into the prokaryotic or eukaryotic cells are described briefly.

Keywords OMVs · Packaging · Interactions · Delivery · Prokaryotes · Eukaryotes · Fusion · Lipid raft · Receptor-mediated endocytosis

An important function of the OMVs is to deliver the luminal or surface associated cargo to a target cell, which may be another Gram-positive or Gram-negative bacterial cell or a eukaryotic cell. Several different means of doing this have been reported (Fig. 6.1). The interaction mechanism may differ with the nature of the target cell. The fusion of OMV (originating from a Gram-negative bacterium) with the outer membrane of another Gram-negative bacterium is quite conceivable since their bounding membranes are similar. But fusion with another Gram-positive bacterial outer membrane is difficult to imagine since the structures of the respective outer membranes are different. In such cases, the other possibility is the lysis of the OMVs at the proximity of the Gram-positive bacteria's outer surface or at some distance there by enabling the cargo to be released and to diffuse through the membranes of the Gram-positive bacteria. Many Gram-negative bacteria contain adhesins on their surface. Accordingly, the OMVs derived from such bacteria are also likely to contain adhesins which enable the OMVs to adhere to the surface of other cells. Interaction between the OMVs and a eukaryotic cell may be imagined to take place in identical ways except that the fusion of the respective membranes is unlikely to occur or difficult to explain. Further, in case of the eukaryotic cell as the target cell, receptor-mediated endocytosis of the entire OMV may take place leading to delivery of the cargo in the cytosol of the target cell. The toxin, such as LT of

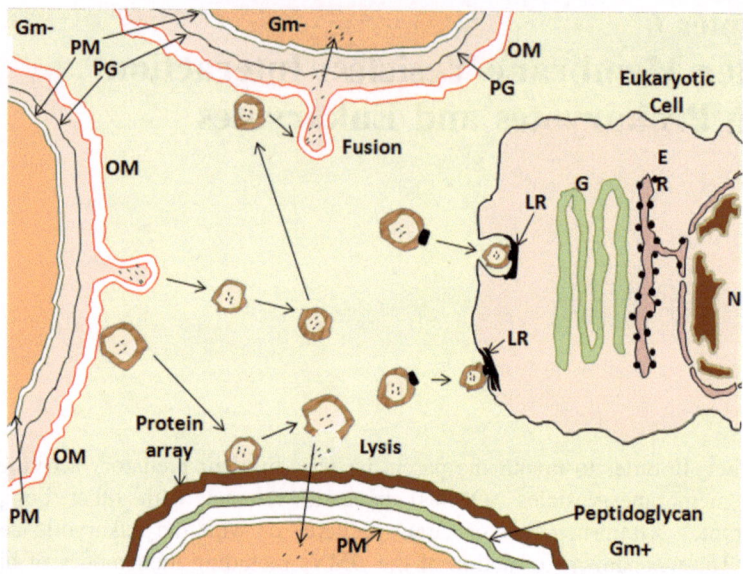

Fig. 6.1 Illustration of the different modes of transfer of the luminal contents or surface associated toxins of OMVs derived from Gram-negative bacterium into (1) another Gram-negative bacterium by fusion with its OM, (2) a Gram-positive bacterium presumably by proximal lysis of the OMVs leading to free diffusion of its contents through the envelope of the host bacterium and (3) an eukaryotic cell by utilizing the toxin receptor and LR mediated endocytosis. *Arrows* show the directions of movement of OMVs for making contact with and transfer of materials into the different host cells. *Gm-*, Gram-negative bacterium; *Gm+*, Gram-positive bacterium, *LR*, lipid raft of the eukaryotic cell membrane; *PM*, plasma membrane; *PG*, peptidoglycan layer; *OM*, outer membrane

enterotoxigenic *Escherichia coli*, gets associated with the LPS on the surface of the vesicles and serves the purpose of adhesin. The interaction between this toxin and receptor on the surface of the eukaryotic cell leads to uptake of the vesicle into the host cell.

The delivery of the virulence factors (toxins and non-toxins) into the target cells via the OMVs may serve different purposes which, in short, may be friendly or hostile. These different purposes, which are backed by experimental evidences, include : (1) survival of the parent bacteria in a mixed population infection by eliminating the competing bacterial strains, (2) protection of the parent bacteria from the co-colonizing bacteria by destroying the toxic factors released by the other bacteria, (3) better growth of the parent bacteria by nutrient acquisition where there is limited nutrition available, (4) increased survival of the parent bacteria through quorum sensing by production and delivery of a sensing molecule which can inactivate the target bacteria and (5) promotion of growth of the co-colonizing bacteria by transferring beneficial materials through OMVs. The ways and means of doing the interactions between the parent bacteria and the target cells along with the purposes served, as recorded in different experimental studies, are discussed in the following.

The OMVs can package the periplasmic peptidoglycan hydrolases (Kadurugamuwa and Beveridge 1995; Li et al. 1996). These enzymes in OMVs have been attributed to the killing of other Gram-negative and Gram-positive bacteria that may be cultured along with (Li et al. 1998), thereby giving to the OMV-producing bacteria an advantage for growth in a mixed bacterial population either in a culture medium or in a host tissue where there is limited nutrition (Ciofu et al. 2000). On the other hand, the OMVs containing β-lactamase can protect the parent bacteria from the co-colonizing β-lactum-producing species that could otherwise eliminate them. In fact, OMVs of *Pseudomonas aeruginosa* strains were found to package β-lactamase produced by these bacteria (Ciofu et al. 2000). Furthermore, some bacteria were provided survival advantage by the OMVs containing scavenging proteases (Dutson et al. 1971; Thompson et al. 1985), xylanase and cellulose (Forsberg et al. 1981), which can help in nutrient acquisition. Gentamicin-induced *P. aeruginosa* OMVs can deliver the antibiotic into the target cell and become bacteriolytic (Kadurugamuwa and Beveridge 1996; Allan and Beveridge 2003). Protease- and toxin- containing vesicles from *E. coli*, *Shigella*, *Actinobacillus*, *Pseudomonas* and *Borrelia* can interact with other bacteria in a way to gain growth advantage over them (Kadurugamuwa and Beveridge 1998, 1999; Kato et al. 2002). These OMVs can interact with bacteria possibly via a fusion or adherence mechanism (Gankema et al. 1980; Kadurugamuwa and Beveridge 1998, 1999; Kato et al. 2002; Saunders et al. 1999; Shoberg and Thomas 1993). Vesicle fusion delivers proteins, such as the autolysin (peptidoglycan hydrolase), that can lyse other ("non-self") Gram-positive and Gram-negative bacteria (Kadurugamuwa and Beveridge 1998; Li et al. 1998). In Gram-positive bacteria, such as *Bacillus stearothermophillus* the peptidoglycan hydrolase penetrates through the S-layer (proteinous layer) of the cell wall and digest the underlying peptidoglycan layer. The S-layer is not affected by attack with OMVs and SDS–polyacrylamide gel electrophoresis showed that the S-layer proteins remained intact. The meso-diaminopimelic acid was found as a breakdown product of peptidoglycan layer in the culture supernatant after the lytic action of OMVs or rather of the autolysin of the OMVs (Kadurugamuwa et al. 1998) was allowed to operate.

The quorum sensing (QS) molecules are synthesized by the bacteria in a population to sense cell density and to perform accordingly group behaviours. Both Gram-negative and Gram-positive bacteria produce these QS molecules, although these molecules may be different structurally and functionally from bacteria to bacteria and even one particular bacterium may contain or produce more than one type of QS molecules. These QS molecules are generally secreted by the bacteria by different mechanisms into the extra cellular milieu. After they reach a critical threshold density, they enter the cell where they bind and activate a transcriptional regulator, leading to significant changes in gene expression. These molecules can even positively regulate gene expression in the parent bacteria as áutoinducers. Although different bacteria produce different types of QS molecules, Gram-negative bacteria produce, in general, the acyl homoserine lactones (HSL) for quorum sensing. The HSLs of all Gram-negative bacteria contain the lactone ring in common, but the acyl chains may vary in lengths and substitutions. However, *P. aeruginosa* produces another cell signalling molecule known as PQS or Pseudomonas Quinolone Signal having the chemical structure as shown in Fig. 6.3 or as 2-heptyl-3-hydroxy-4-quinolone (Pesci et al. 1999). It was found that the PQS molecules were secreted by the *P. aeruginosa* via the OMVs and that they formed an important virulence component of the vesicles. These *Pseudomonas* OMVs can participate in QS and thereby aid in communication within mixed populations of pathogens leading to their increased survival. Further, the PQS molecules can stimulate the production of OMVs by the *Pseudomonas* (Mashburn and Whiteley 2005), a property that will be discussed further elsewhere.

The OMVs can also transfer beneficial materials and promote the growth of co-colonizing pathogens. In a mixed population of bacteria, the OMVs that can titrate an antimicrobial agent or degrade *β*-lactums significantly promote the survival of any neighbouring bacteria. Again, the OMVs released by one strain can cause increased inflammation leading to the exposure of host extracellular matrix proteins and upregulation of the epithelial cell surface receptors that are beneficial to colonization by another bacterial strain. It was proposed that OMV-mediated pro-inflammatory changes in host tissue can help in the adherence and survival of co-colonizing pathogens. Thus *Moraxella catarrhalis* is often found in mixed infections with pathogens like *Hemophilus influenzae*.

6.1 Binding with Target Cells Membrane

A number of evidences show that OMVs produced by a variety of Gram-negative bacteria bind to cultured host cells. This binding may be directly receptor-mediated or receptor-mediated after the receptor is uncovered by an enzyme of the OMVs (Bauman and Kuehn 2009). Thus the OMV-associated aminopeptidase was shown to increase the ability of *P. aeruginosa* vesicles to bind to primary and also cultured human lung epithelial cells (Bauman and Kuehn 2009). The OMVs derived from several bacteria, viz., *E. coli, Shigella, Actinobacillus, Borrelia* strains etc., and containing protease and toxins were shown to bind to bacterial as well as mammalian cell membranes (Gankema et al. 1980; Kadurugamuwa and Beveridge 1998; Kato et al. 2002; Saunders et al. 1999; Shoberg and Thomas 1993). OMVs were also shown to bind to membranes of cells within the infected tissue. OMVs from *Helicobacter pylori* were shown to bind to intestinal cells in specimens biopsied from infected patients (Heczko et al. 2000; Keenan et al. 2000). Binding of OMVs to host cells depends on factors specific to the parent bacteria. The OMVs and the parent bacteria can use the identical host cell receptors there by giving rise to a competition between them. Thus, OMVs from *Borrelia burgdorferi* were found to compete with the corresponding whole *Borrelia* cells for binding to human umbilical vein endothelial cells (Whitmire and Garon 1993).

6.2 Fusion with Target Cell Membrane

Although there are evidences showing fusion of the OMVs with the host eukaryotic cell plasma membrane, the mechanism how this could take place remains unclear (Ellis and Kuehn 2010; Kuehn and Kesty 2005). This question arises particularly because the two membranes, the membrane bounding the OMVs and the plasma membrane of the eukaryotic cell, are so different in structure and chemical composition. It has been reported that the *Salmonella* cells growing within the vacuoles of host cells released LPS containing O-antigen and that these LPSs were found within the epithelial cell vesicle membranes, as detected by the antibodies to the O antigens (Garcia-del Portillo et al. 1997). Similarly, the vesicles derived from *Actinobacillus actinomycetemcomitans* were able to introduce a lipid-tracking dye to host cell plasma membranes within a short time after co-incubation (Demuth et al. 2003). Again, OMVs derived from *Legionella pneumophila* bearing a fluorescent label were found either associated for over a long period of time or fused with the alveolar epithelial cell membrane (Galka et al. 2008). Again, such fusion of the bacterial structures with the epithelial cell membrane has to be explained properly particularly since it may have significant impact on the host's immunological responses.

6.3 Entry into the Target Cells

Entry of OMVs into the target cells may be either by the receptor-mediated endocytic pathway or by other means not very clearly defined. The best example, as discussed before, of the receptor-mediated endocytic entry is that of the enterotoxin, LT produced by enterotoxigenic *E. coli* cells (ETEC). The toxin molecule, LT contains one A subunit and five B subunits and LT is released by the ETEC cells in association with the OMVs and are bound to the outer surface LPS of the OMVs. The OMV.LT complex then migrates toward the epithelial cell membrane where binding through LT takes place to the G_{M1} sites and the lipid raft. Then the lipid raft (Fig. 6.2a, b) mediated endocytosis of the OMV.LT complex takes place. Once inside the cell, the A subunit of LT migrates to the endoplasmic reticulum where it undergoes conformational change and is released to the cytosol (Fig. 6.3). Similarly, a fraction of CT has recently been shown to be associated with the LPS of *Vibrio cholerae* OMVs (Chatterjee and Chaudhuri 2011). These OMV.CT complexes then get entry into the epithelial cells exactly in the same way as the OMV.LT complexes.

VacA of *H. pylori* is released into the extra cellular medium both as a free toxin and as an OMV complex. Both forms of VacA bind to the gastric epithelial cell surface (Fiocca et al. 1999; Ricci et al. 2005). Whether VacA acts as the adhesin for the OMV or VacA complex binds to the epithelial cell surface remains to be elucidated. The protein, OmpA, the common component of the outer membrane of

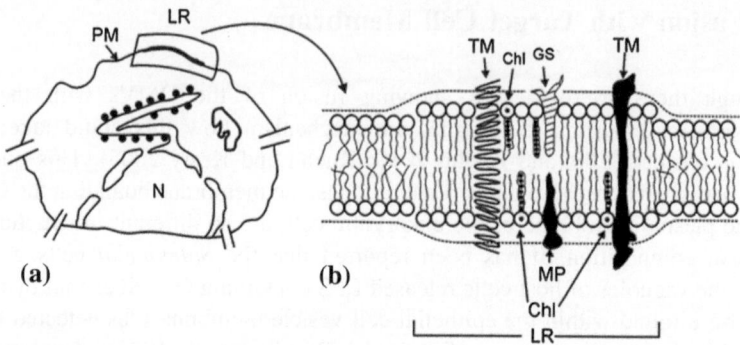

Fig. 6.2 The presence of LR in the plasma membrane (*PM*) of the epithelial cell. **a** Part of a cell showing the plasma membrane (*PM*), nucleus (*N*), portion of PM containing LR etc. **b** Enlarged view of the portion of the plasma membrane within the box in 'A' showing the organization of the PM containing lipid LR. The LR is a specialised membrane domain containing high concentrations of cholesterol (Chl), sphingomyelin, gangliosides (*GS*), etc. This region is also enriched in phospholipids that contain saturated fatty acyl chains (straight lines in lipid tails). A variety of proteins- membrane proteins (*MP*) and transmembrane proteins(*TM*)-partition into LR. The composition of LR results in lateral phase separation and generation of a liquid-ordered domain [from (Chaudhuri and Chatterjee 2009)]

the parent bacteria and the vesicles derived therefrom was also shown to contribute to the entry of vesicles into the host cells. *E. coli* K1, the causative agent of the neonatal meningitis, was shown to undergo OMV-mediated entry into the brain microvascular endothelial cells through an interaction between OMP and Ecgp, a surface receptor on the brain cells (Prasadarao 2002; Prasadarao et al. 1996). However, OMV binding to host cell may not always be mediated by a toxin and OMVs derived from *A. actinomycetemcomitans* do not require leukotoxin for its interaction with the host cell (Demuth et al. 2003), although such OMVs are rich in leukotoxin content (Kato et al. 2002). Host factors were also shown to mediate OMVs entry into specific host cells. It was found that the bactericidal/permeability increasing protein (BPI) and not the LPS binding protein (LBP) contributed to the increased entry of OMVs derived from *N. meningitides* into the dendritic cells. In fact, the BPI co-localized with the internalized OMVs could be directly visualized by confocal microscopy (Schultz et al. 2007). However, such action of BPI has not been found to be a general phenomenon applicable to interactions between any cell and BPI.

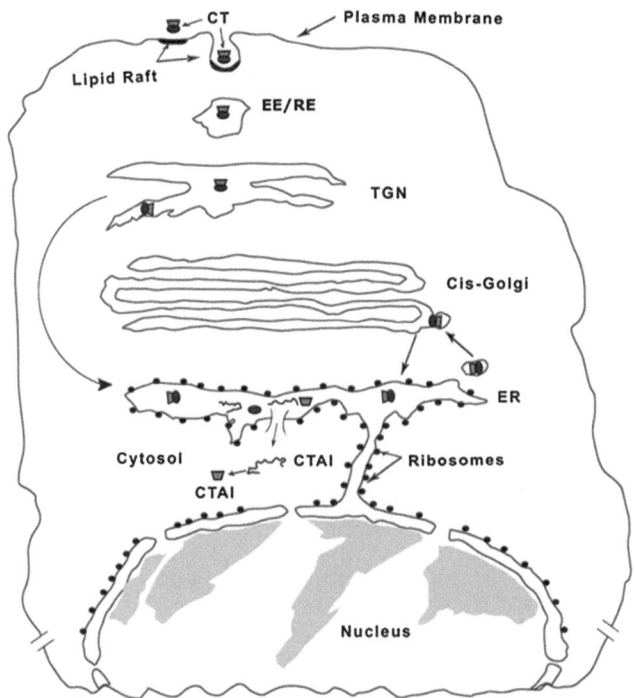

Fig. 6.3 Schematic diagram illustrating the endocytosis of cholera toxin, CT, and its retrograde trafficking via the trans-golgi network (*TGN*) to the lumen of the endoplasmic reticulum (*ER*), bypassing the Cis-Golgi complex (Cis-Golgi). Some of the CT molecules in the ER lumen may travel backward to the Cis-Golgi complex again return to the lumen of ER (shown by *arrows*) in a KDEL-mediated retrieval step. In the ER lumen, CT undergoes disassembly and the dissociated CTA1 fragment gets unfolded and translocated across the Sec 61 channel of the ER membrane to the cytosol of the cell, where it gets quickly refolded to its native structure (shown by *arrow*) [from (Chaudhuri and Chatterjee 2009)]

References

Allan ND, Beveridge TJ (2003) Gentamicin delivery to *Burkholderia cepacia* group IIIa strains via membrane vesicles from *Pseudomonas aeruginosa* PAO1. Antimicrob Agents Chemother 47:2962–2965

Bauman SJ, Kuehn MJ (2009) *Pseudomonas aeruginosa* vesicles associate with and are internalized by human lung epithelial cells. BMC Microbiol 9:26

Chatterjee D, Chaudhuri K (2011) Association of cholera toxin with *Vibrio cholerae* outer membrane vesicles which are internalized by human intestinal epithelial cells. FEBS Lett 585:1357–1362

Chaudhuri K, Chatterjee SN (2009) Cholera toxins. Springer, Berlin

Ciofu O, Beveridge TJ, Kadurugamuwa J, Walther-Rasmussen J, Hoiby N (2000) Chromosomal beta-lactamase is packaged into membrane vesicles and secreted from *Pseudomonas aeruginosa*. J Antimicrob Chemother 45:9–13

Demuth DR, James D, Kowashi Y, Kato S (2003) Interaction of *Actinobacillus actinomycetemcomitans* outer membrane vesicles with HL60 cells does not require leukotoxin. Cell Microbiol 5:111–121

Dutson TR, Pearson AM, Price JF, Spink GC, Tarrant PJ (1971) Observations by electron microscopy on pig muscle inoculated and incubated with *Pseudomonas fragi*. Appl Microbiol 22:1152–1158

Ellis TN, Kuehn MJ (2010) Virulence and immunomodulatory roles of bacterial outer membrane vesicles. Microbiol Mol Biol Rev 74:81–94

Fiocca R, Necchi V, Sommi P, Ricci V, Telford J, Cover TL, Solcia E (1999) Release of *Helicobacter pylori* vacuolating cytotoxin by both a specific secretion pathway and budding of outer membrane vesicles. Uptake of released toxin and vesicles by gastric epithelium. J Pathol 188:220–226

Forsberg CW, Beveridge TJ, Hellstrom A (1981) Cellulase and xylanase release from *Bacteroides succinogenes* and its importance in the rumen environment. Appl Environ Microbiol 42:886–896

Galka F, Wai SN, Kusch H, Engelmann S, Hecker M et al (2008) Proteomic characterization of the whole secretome of *Legionella pneumophila* and functional analysis of outer membrane vesicles. Infect Immun 76:1825–1836

Gankema H, Wensink J, Guinee PA, Jansen WH, Witholt B (1980) Some characteristics of the outer membrane material released by growing enterotoxigenic *Escherichia coli*. Infect Immun 29:704–713

Garcia-del Portillo F, Stein MA, Finlay BB (1997) Release of lipopolysaccharide from intracellular compartments containing *Salmonella typhimurium* to vesicles of the host epithelial cell. Infect Immun 65:24–34

Heczko U, Smith VC, Mark Meloche R, Buchan AM, Finlay BB (2000) Characteristics of *Helicobacter pylori* attachment to human primary antral epithelial cells. Microbes Infect 2:1669–1676

Kadurugamuwa JL, Beveridge TJ (1995) Virulence factors are released from *Pseudomonas aeruginosa* in association with membrane vesicles during normal growth and exposure to gentamicin: a novel mechanism of enzyme secretion. J Bacteriol 177:3998–4008

Kadurugamuwa JL, Beveridge TJ (1996) Bacteriolytic effect of membrane vesicles from *Pseudomonas aeruginosa* on other bacteria including pathogens: conceptually new antibiotics. J Bacteriol 178:2767–2774

Kadurugamuwa JL, Beveridge TJ (1998) Delivery of the non-membrane-permeative antibiotic gentamicin into mammalian cells by using *Shigella flexneri* membrane vesicles. Antimicrob Agents Chemother 42:1476–1483

Kadurugamuwa JL, Beveridge TJ (1999) Membrane vesicles derived from *Pseudomonas aeruginosa* and *Shigella flexneri* can be integrated into the surfaces of other gram-negative bacteria. Microbiol 145(Pt 8):2051–2060

Kadurugamuwa JL, Mayer A, Messner P, Sara M, Sleytr UB, Beveridge TJ (1998) S-layered *Aneurinibacillus and Bacillus* spp. are susceptible to the lytic action of *Pseudomonas aeruginosa* membrane vesicles. J Bacteriol 180:2306–2311

Kato S, Kowashi Y, Demuth DR (2002) Outer membrane-like vesicles secreted by *Actinobacillus actinomycetemcomitans* are enriched in leukotoxin. Microb Pathog 32:1–13

Keenan J, Day T, Neal S, Cook B, Perez–Perez G, Allardyce R, Bagshaw P (2000) A role for the bacterial outer membrane in the pathogenesis of *Helicobacter pylori* infection. FEMS Microbiol Lett 182:259–264

Kuehn MJ, Kesty NC (2005) Bacterial outer membrane vesicles and the host-pathogen interaction. Genes Dev 19:2645–2655

Li Z, Clarke AJ, Beveridge TJ (1996) A major autolysin of *Pseudomonas aeruginosa*: subcellular distribution, potential role in cell growth and division and secretion in surface membrane vesicles. J Bacteriol 178:2479–2488

Li Z, Clarke AJ, Beveridge TJ (1998) Gram-negative bacteria produce membrane vesicles which are capable of killing other bacteria. J Bacteriol 180:5478–5483

Mashburn LM, Whiteley M (2005) Membrane vesicles traffic signals and facilitate group activities in a prokaryote. Nature 437:422–425

Pesci EC, Milbank JB, Pearson JP, McKnight S, Kende AS, Greenberg EP, Iglewski BH (1999) Quinolone signaling in the cell-to-cell communication system of *Pseudomonas aeruginosa*. Proc Natl Acad Sci U S A 96:11229–11234

Prasadarao NV (2002) Identification of *Escherichia coli* outer membrane protein A receptor on human brain microvascular endothelial cells. Infect Immun 70:4556–4563

Prasadarao NV, Wass CA, Weiser JN, Stins MF, Huang SH, Kim KS (1996) Outer membrane protein A of *Escherichia coli* contributes to invasion of brain microvascular endothelial cells. Infect Immun 64:146–153

Ricci V, Chiozzi V, Necchi V, Oldani A, Romano M, Solcia E, Ventura U (2005) Free-soluble and outer membrane vesicle-associated VacA from *Helicobacter pylori*: two forms of release, a different activity. Biochem Biophys Res Commun 337:173–178

Saunders NB, Shoemaker DR, Brandt BL, Moran EE, Larsen T, Zollinger WD (1999) Immunogenicity of intranasally administered meningococcal native outer membrane vesicles in mice. Infect Immun 67:113–119

Schultz H, Hume J, de Zhang S, Gioannini TL, Weiss JP (2007) A novel role for the bactericidal/permeability increasing protein in interactions of Gram-negative bacterial outer membrane blebs with dendritic cells. J Immunol 179:2477–2484

Shoberg RJ, Thomas DD (1993) Specific adherence of *Borrelia burgdorferi* extracellular vesicles to human endothelial cells in culture. Infect Immun 61:3892–3900

Thompson SS, Naidu YM, Pestka JJ (1985) Ultrastructural localization of an extracellular protease in *Pseudomonas fragi* by using the peroxidase-antiperoxidase reaction. Appl Environ Microbiol 50:1038–1042

Whitmire WM, Garon CF (1993) Specific and nonspecific responses of murine B cells to membrane blebs of *Borrelia burgdorferi*. Infect Immun 61:1460–1467

Chapter 7
Biogenesis of Outer Membrane Vesicles

Abstract The different steps involved in the formation and release of outer membrane vesicles (OMVs) from the parent bacteria as revealed by the experimental evidence and derived models presented by different investigators are discussed. An attempt has been made to develop a unified concept of the biogenesis of OMVs released by different Gram-negative bacteria.

Keywords OMVs · Models of biogenesis · Sorting processes · Curvature protein · PQS

Bacteria growing under many different conditions produce OMVs. The different growth media include solid agar (Tetz et al. 1990), different liquid media (Devoe and Gilchrist 1973; Hozbor et al. 1999; Shoberg and Thomas 1995; Wai et al. 1995), biofilms (Schooling and Beveridge 2006; Yonezawa et al. 2009), intracellular infections (Galka et al. 2008), and in vivo (Brandtzaeg et al. 1992; Hynes et al. 2005; Stephens et al. 1982; Tan et al. 2007). However, all these media are not equally favorable for OMV production by bacteria. The presence of stress in growth has often been found to increase OMV production. These stress factors include (1) the presence of antibiotics such as gentamicin (Kadurugamuwa and Beveridge 1995) and chloramphenicol (Rothfield and Pearlman-Kothencz 1969) and the like, (2) limitation in growth media of various nutrients such as lysine (Bishop and Work 1965; Knox et al. 1966), Mg^{2+} (Suzuki et al. 1978) or hemin (Smalley et al. 1991), (3) phage infections (Loeb 1974), and (4) presence of unfolded proteins in the periplasm causing stress to envelope and lead to overproduction of the OMVs (McBroom and Kuehn 2007) and so on. Under some of these stress conditions, such as growth of lysine-requiring *E. coli* in lysine-deficient media (Knox et al. 1966) or growth under treatment of gentamicin (Kadurugamuwa and Beveridge 1995), the OMVs produced are different from those produced naturally,that is, during active growth of bacteria in nutritionally rich media in that they contain cytosolic components and may be leaky and are

S. N. Chatterjee and K. Chaudhuri, *Outer Membrane Vesicles of Bacteria*,
SpringerBriefs in Microbiology, DOI: 10.1007/978-3-642-30526-9_7,
© The Author(s) 2012

now termed unnatural OMVs. They are not produced by a well-regulated cellular process but by cells undergoing death or decay (under S.O.S) and with abnormal changes in structure. While considering the biogenesis of OMVs, only the naturally produced or natural OMVs are considered here.

It is believed that the first step in vesicle production lies in the detachment of the outer membrane (OM) from the peptidoglycan layer at regions where there is faster growth of the OM (Chatterjee and Das 1967; Wensink and Witholt 1981) *vis-à-vis* the peptidoglycan layer and the inner membrane or where fewer linkages between the OM and the peptidoglycan layer or the inner membrane are present. A different view proposed that there are regions in the periplasm where an imbalance in the turnover of PG leads to an accumulation of muramic acid, which generated turgor pressure and caused a bulging of the OM (Zhou et al. 1998). In fact, some OMVs were found to contain low molecular weight muramic acid (Kadurugamuwa and Beveridge 1995; Work et al. 1966; Zhou et al. 1998). On the other hand, charge–charge interaction might play a role in the causation of OMVs. The OM of *Pseudomonas aeruginosa* is enriched with the negatively charged B-band LPS, which generates a strong repellent force there by leading to a higher rate of OMV production (Kadurugamuwa and Beveridge 1995; Nguyen et al. 2003; Sabra et al. 2003). Without overruling any of these views or findings, the facts that emerged more recently showed that there are localized regions in the OM where there is significant absence of OM proteins covalently or non covalently bound to the peptidoglycan layer protein or proteins of the inner membrane or other membrane proteins leading to the production of OMVs from these regions. This is consistent with the finding that the OMVs lacked some lipoproteins compared to the OM (Hoekstra et al. 1976; Wensink and Witholt 1981). Independent studies revealed that in several bacterial spp. such as *E. coli* (McBroom et al. 2006; Sonntag et al. 1978), *Porphyromonas gingivalis* (Iwami et al. 2007), *Salmonella typhimurium* (Deatherage et al. 2009), and the like, the proteins OmpA, LppA/B, and TolA/B form the major membrane proteins that stabilize the OM through interactions with the proteins of the inner membrane or those of the peptidoglycan layer. Deletion of these was shown to increase the OMV production rate. In fact, a reduction in the number of cross-links between the PG and the OM may govern sites of OMV production (Hoekstra et al. 1976).

Genetic studies are expected to shed light on the roles of different membrane proteins in the formation of the OMVs. It has been suggested from genetic evidence that OMV production cannot be abolished (McBroom et al. 2006; McBroom and Kuehn 2005). McBroom et al. (2006) conducted a screen in *E. coli* to identify gene disruptions that caused vesicle over- or under-production and revealed that vesiculation is a fundamental characteristic of Gram-negative bacterial growth and is not a consequence of bacterial lysis or disintegration of the bacterial envelope. Gene disruptions were identified that caused differences in vesicle production ranging from a five-fold decrease to a 200-fold increase relative to wild-type levels. These gene disruptions were identified in loci controlling OM components, peptidoglycan synthesis, and even the σ^E cell envelope stress response. Their study on detergent sensitivity, leakiness, and growth characteristics of the vesiculation mutant strains demonstrated that vesicle production was not predictive of envelope instability. The lipoprotein Lpp contributes significantly to the interaction between the OM and the peptidoglycan layer and a mutation in the corresponding gene was found to cause hypervesiculation (Bernadac et al. 1998; Cascales et al. 2002). The gene *yfgL* encodes a lipoprotein involved in the synthesis and/or degradation of the peptidoglycan layer (Eggert et al. 2001). A deletion of this gene caused a significant decrease in vesiculation. It was proposed that the role of *yfgL* in vesiculation was the result of an increase in the production of the peptidoglycan or a downregulation of the lytic transglycosylases, which leads to a loss of turgor pressure in the OM (Rolhion et al. 2005). A pronounced OM defect was observed with *lpp-ompA* cells, which form OMVs under normal growth conditions (Sonntag et al. 1978). *lpp* strains producing extra amounts of periplasmic ß-lactamase also produced OMVs (Bernadac et al. 1987). Furthermore, the *lpp* strains in which the link between the outer membrane and the peptidoglycan was lacking also produced OMVs upon Mg^{2+} starvation (Suzuki et al. 1978; Yem and Wu 1978); Bernadac et al. (1987) used Western blotting and electron microscopic techniques to study several *E. coli* strains bearing defects in the OM integrity and found that mutations in each of the *tol-pal* genes led to the formation of OMVs. A *tolA* deletion in three different *E. coli* strains led to production of higher amounts of OMVs. The authors concluded that defects in any of the Tol-Pal proteins led to the formation of OMVs. In a more recent study, Song et al. (2008) detected in *V. cholerae* O1 strain A1552 a new small s-RNA gene (140 nt), *vrrA*, whose expression required the membrane stress factor σ^E. It was suggested that VrrA acted on *ompA* in response to periplasmic protein folding stress. The authors further observed that OmpA levels inversely correlated with the number of OMVs produced and that VrrA increased OMV production comparable to loss of OmpA. Thus, VrrA positively controlled the release of OMVs by negatively controlling the expression of OM protein OmpA.

Deatherage et al. (2009) adopted a quantitative approach to study the mechanism of OMV formation in wild-type (WT) and mutants of *Salmonella* strains and revealed the importance of protein–protein interactions in the cell envelope. In the WT cells the major OMV proteins identified were OmpC, OmpF, NmpC, OmpX, OmpA, LppAB, Pal, and TolB. The mutants lacking OmpC, OmpF, NmpC, or OmpX produced OMVs similar to WT cells, whereas those lacking OmpA, LppAB, Pal, or TolA significantly increased OMV production. The size distribution of the OMVs also depended on the nature of mutations. The *pal*, *tolB*, and *tolA* mutants released OMVs significantly larger than WT cells. The authors (Deatherage et al. 2009) proposed a model of OMV biogenesis, wherein bacterial growth and division invoked temporary localized reductions in the density of OM-PG and OM-PG-IM associations within the envelope structure leading to release of OMVs from those areas. The genes of the *tol/pal* system were considered to be a major regulator of OMV formation by Gram-negative bacteria (Bernadac et al. 1998; Deatherage et al. 2009; Henry et al. 2004). But defects in this system were found to cause significant leakiness of the OM (Lazzaroni et al.

1999; Llamas et al. 2000) unlike some other mutations that caused vesiculation without any defect in the OM (McBroom et al. 2006). Two hypervesiculation mutants were identified by performing a transposon mutant screen of *E. coli* having mutations in the genes *ypjA* and *nlpA*, which encoded putative cell enve-lope-localized proteins. Several other hypervesiculating mutants were identified that produced increasing amounts of OM proteins and lipids extracellularly, presumably because of damage or disintegration of the OM during vesiculation (McBroom et al. 2006).

The role of the stress response σ^E pathway in producing OMVs is of consider-able importance. Transposon mutations in several genes of this pathway caused vesiculation with practically no OM damage. In general, vesiculation increased when the σ^E pathway was activated (McBroom and Kuehn 2007). However, hypervesiculation mutants with significantly lower σ^E activities were also identified (McBroom and Kuehn 2007). This might be because of accumulation of consid-erable amounts of misfolded proteins in the periplasm causing increased pressure on the OM. Several results indicated that accumulation of excess amounts of non-native or native proteins in the periplasm led to a significant increase of vesiculation in *E. coli* (McBroom and Kuehn 2007). This effect was dose-depen-dent, not dependent solely on endogenous proteins and not due to σ^E activation and was thus due to overabundance of periplasmic materials. A role for the OM–peptidoglycan links in vesiculation has been evident from many studies. OmpA being an important protein linking the OM and the peptidoglycan layer, its deletion or truncation was shown to increase vesiculation in *Salmonella, E. coli,* and *V. cholerae cells* (Deatherage et al. 2009; Song et al. 2008; Sonntag et al. 1978). A study using truncation mutants of some other abundant OM–peptidoglycan-linking proteins in *Salmonella spp* (Deatherage et al. 2009) supported the role of OM–peptidoglycan links in the vesiculation process. Vesiculation was shown to increase to different extents by truncation. This strengthened the idea that OM–peptidoglycan links play an important role in regulating the vesiculation process.

Proteomic analyses of the OMVs may help in our understanding of their biogenesis. The outer membrane proteins, Omps, Tol-Pal, YbgF, and Lpp found in the OMV proteome may play a role in the liberation of OMVs from the bacterial surface by producing faster expansion or growth of the OM *vis-à-vis* the pepti-doglycan layer (Bernadac et al. 1998), although the actual mechanism of the function of these proteins remains largely unclear. It is further believed that the other proteins, murein hydrolases (including MltA, MipA, MltE, and SLP) found in OMVs, may produce peptidoglycan fragments in the periplasm leading to increased turgor pressure and discharge of the OMVs (Lommatzsch et al. 1997). Proteome analyses have shown that although the outer membrane and periplasmic proteins are highly enriched in the OMVs, the inner membrane and the cyto-plasmic proteins are practically excluded (Lee et al. 2007). The EchoBASE database shows that among the total number of 4345 *E. coli* proteins (Table 7.1) the OM and periplasmic proteins are more commonly sorted into OMVs (Misra et al. 2005). This is more evident from the data of column 5, Table 7.1, showing

Table 7.1 Relative distribution of proteins in E.coli whole bacteria and OMVs[a]

Cellular compartments	Whole cell		Outer membrane vesicle		
	Number of proteins	Fraction of A (%)	Number of proteins of Col.2	Fraction of corresponding Proteins in Col.2 (%)	Fraction of B (%)
Extra-cellular space	10	0.2	1.0	10.0 (?)	0.7
Outer membrane (OM)	149	3.4	65.0	43.6	46.1
Periplasm (PPM)	350	8.1	16.0	4.57	11.3
Inner membrane (IM)	974	22.4	7.0	0.7	5.0
Cytoplasm (CP)	2862	65.9	52.0	1.80	36.9 (?)
Total	4345 (A)	100	141 (B)		100

[a] Based on data presented in Lee et al. (2007, 2008) and Post et al. (2005).

that the OM proteins are mostly sorted into the OMVs (43.6 %) and then the periplasmic proteins (4.57 %). The inner membrane and the cytoplasmic proteins are not significantly sorted into the OMVs. However, the data in column 5, Table 7.1, show that proteins from extra-cellular space are also sorted out into the OMVs (10 %), which need to be explained properly. The data presented in column 6, Table 7.1, also support the fact that the proteins of the OM and the periplasm are sorted out into the OMVs. But the presence of a significant amount of cytoplasmic proteins in the OMVs (36.9 %), as shown in column 6, Table 7.1, needs to be reinvestigated. Analyses of proteomic data further show that the inclusion of any particular protein into the OMVs is not in accordance with its abundance in the whole bacterial cell (Corbin et al. 2003), supporting the hypothesis that there is a special sorting mechanism in action during the OMV formation or that the OMVs are formed from specific sites of the bacterial surface, which, however, is against our common experience.

7.1 Model of OMV Biogenesis

Several models or ideas have been put forward to explain the biogenesis of OMVs of different bacteria. Before considering any one or all of these models or ideas it would be worthwhile to reconsider or review briefly the findings of different authors relevant to building a general idea about the biogenesis of OMVs. Also it is better to separate the cases where OMVs or OMV-like products are produced under strong challenging stresses making the very survival of the bacteria concerned at stake, that is, under unnatural conditions. Under such conditions the bacteria are not expected to exhibit any well-regulated cellular activity but act desperately for survival. Such things may lead to widespread breakage of the peptidoglycan layer by treatment with specific antibiotics or autolysins or during abnormal growth in a particular nutrientless or deficient medium, and so on. It is therefore futile to develop a generalized model of biogenesis that will also cover those abnormal conditions. In fact, biogenesis of OMVs has now been acknowledged as an essential, well-regulated, and possibly well-conserved process (Kulp and Kuehn 2010).

An early step or perhaps the first step in OMV production is localized bulging of the OM at a few or more regions of the OM. This requires (1) production of extra amounts of the OM components in order to avoid local disruptions of the OM, (2) local changes in the association between the OM and the peptidoglycan layer which could be possible either by the breakage of the OM and the pepti-doglycan protein bonds or by lateral shifting of the interacting OM and the PG proteins from the concerned area(s), (3) a supply of energy for the process of bulging to take place, and (4) the cell's motive or purpose for invoking such bleb formation. Any model for bleb formation is expected to take into account all the factors referred to above. It has been recognized that vesiculation can increase even a hundred fold or more without affecting membrane integrity (McBroom et al. 2006), which is practically impossible if the peptidoglycan layer is disinte-grated (Kulp and Kuehn 2010).

It is rather universally accepted that for local bulging(s) to take place the synthesis of an extra amount of OM components is a must. There is enough evidence to support the second factor: the OM peptidoglycan links could be shifted from the bulging site or subjected to degradation or destruction. Several investi-gators have shown that the linking protein OmpA is depleted but present in OMVs and Lpp is excluded (Hoekstra et al. 1976; Loeb and Kilner 1978; Mug-Opstelten and Witholt 1978; Wensink and Witholt 1981). Also, it is widely acknowledged that shifting or relocating the OM-PG links have the advantage that the membrane integrity can be maintained at all levels of vesiculation. However, any direct evidence of such shifting or relocation of the links remains to be shown. As for the supply of energy there is some degree of vagueness in the experimental evidence presented thus far. One idea is that assembly of different proteins, fully folded or unfolded, in the periplasmic space between the OM and the PG layer produce turgor pressure on the OM thereby causing its outward bulging. Detection of OMVs enriched with such types of proteins lend credence to this idea (Horstman and Kuehn 2000; Kato et al. 2002; McBroom and Kuehn 2007). In addition, it has been proposed that the bulging of OM could be caused by the presence of a special type of protein in association with the OM in the locality, the so-called curvature-inducing proteins. The idea is that upon gathering of the curvature-inducing molecules in the region, the OM–PG links might be broken or shifted leading to the bulging of OM in that area. Furthermore, the curvature-inducing molecules may lead to enrichment of certain proteins or other molecules in the OMVs. However, no direct evidence of any curvature-inducing molecule taking part in the causation of OMV bulging has yet been reported.

On the other hand, it was shown that a quorum sensing molecule, PQS (Pseudomonas Quinolone Signal) also initiated OMV formation in *P. aeruginosa*. A direct correlation was found between the alkyl chain length and the OMV-forming ability: the longer the chain length the greater was the capacity to induce OMVs (Mashburn-Warren 2008). The author further showed that PQS interacted with the LPS of the OM and that the length of its alkyl chain and the presence of

Table 7.2 Summary of events involved in the biogenesis of the outer membrane vesicles (*OMVs*)

Stages	Activities on the surface of bacterial cell
1	Extra growth of the OM vis-à-vis PG
2	Accumulation of the extra materials, such as PG fragments,unfolded or misfolded proteins, and so on, in the periplasm (PPM) between OM and PG. causing increased turgor pressure
3	Absence or lateral transfer of proteins interacting between OM and PG from the area of bulging
4	Continued growth of OM, accumulation of materials in the PPM, and absence of covalent or noncovalent force between PG and OM leading to an increase in size of the bleb, formation of the constricted neck, and pinching off of the bleb into the extracellular medium

Note: 1. Stages 1–4 do not represent the actual sequence of occurrence of events or activities described in column 2. In fact, there may be considerable over lapping between the timing of activities shown in different stages.
2. The source or sources of energy required for all the activities referred to above remain unclear at the moment.
3. Rate of increase of curvature of the bleb is likely to decide its number and size. The rate may be increased by the presence of any curvature-inducing molecule.

the hydroxyl group on the third carbon (Fig. 4.3) were critical for PQS incorporation into LPS. This incorporation was presumed to cause a more ordered structure of the LPS. It was further proposed that ordering of the LPS structure would lead to curvature formation in the corresponding area of the OM, a process that would initiate the OMV formation. PQS was thus considered to be a molecule producing curvature in the OM of *P. aeruginosa*, of course at those areas of the OM where there are few or no proteins interacting between the peptidoglycan layer and the outer and/or inner membrane. The presence of any similar quorum sensing molecule in other Gram-negative bacteria is yet to be reported or identified.

It thus appears that the presence of any curvature-inducing molecule, other than PQS in *P. aeruginosa*, in the Gram-negative bacterial OM remains to be detected and characterized experimentally. For the present, the fact remains that OMVs are produced even without the presence of PQS in different bacteria and hence the role of any curvature-inducing molecule in the OMV formation may be kept out of consideration. With respect to the supply of energy for production and fission of OMVs from any bacteria, no direct evidence is available thus far. There are different modes of energy transfer from the bacterial cytoplasm to the periplasmic space through the plasma or inner membrane (IM) and one or none of these modes of energy transfer may play a role in the production of OMVs. However, further investigation is required for arriving at any conclusion.

In an elegant review, Kulp and Kuehn proposed three apparently different pathways (which are not mutually exclusive) for biogenesis of OMVs (Kulp and Kuehn 2010). If the role of curvature-inducing proteins is kept out of consideration until any direct evidence of their presence in the vesicles, except the PQS molecule in *P.aeruginosa*, is experimentally established, the first and second pathways depicted by the authors (Kulp and Kuehn 2010) become the same if we assume that the initiation of bulging takes place by the presence or accumulation of proteins or any other molecules that the cell wants to get rid of for different purposes, in the local periplasmic space. Depending on the nature of the cargo, more or less vesiculation may take place. This concept (which is in reality a slight modification of the unification of the first two processes described by Kulp and Kuehn (2010) of the biogenesis of OMVs) is thus briefly presented in Table 7.2. However, the nature and the mechanism of energy transfer in the biogenesis of OMVs still remain unclear.

References

Bernadac A, Bolla JM, Lazdunski C, Inouye M, Pages JM (1987) Precise localization of an overproduced periplasmic protein in *Escherichia coli*: use of double immuno-gold labelling. Biol Cell 61:141–147

Bernadac A, Gavioli M, Lazzaroni JC, Raina S, Lloubes R (1998) *Escherichia coli tol-pal* mutants form outer membrane vesicles. J Bacteriol 180:4872–4878

Bishop DG, Work E (1965) An extracellular glycolipid produced by *Escherichia coli* grown under lysine-limiting conditions. Biochem J 96:567–576

Brandtzaeg P, Bryn K, Kierulf P, Ovstebo R, Namork E, Aase B, Jantzen E (1992) Meningococcal endotoxin in lethal septic shock plasma studied by gas chromatography, mass-spectrometry, ultracentrifugation, and electron microscopy. J Clin Invest 89:816–823

Cascales E, Bernadac A, Gavioli M, Lazzaroni JC, Lloubes R (2002) Pal lipoprotein of *Escherichia coli* plays a major role in outer membrane integrity. J Bacteriol 184:754–759

Chatterjee SN, Das J (1967) Electron microscopic observations on the excretion of cell-wall material by *Vibrio cholerae*. J Gen Microbiol 49:1–11

Corbin RW, Paliy O, Yang F, Shabanowitz J, Platt M et al (2003) Toward a protein profile of *Escherichia coli*: comparison to its transcription profile. Proc Natl Acad Sci USA 100: 9232–9237

Deatherage BL, Lara JC, Bergsbaken T, Rassoulian Barrett SL, Lara S, Cookson BT (2009) Biogenesis of bacterial membrane vesicles. Mol Microbiol 72:1395–1407

Devoe IW, Gilchrist JE (1973) Release of endotoxin in the form of cell wall blebs during in vitro growth of *Neisseria meningitidis*. J Exp Med 138:1156–1167

Eggert US, Ruiz N, Falcone BV, Branstrom AA, Goldman RC, Silhavy TJ, Kahne D (2001) Genetic basis for activity differences between vancomycin and glycolipid derivatives of vancomycin. Science 294:361–364

Galka F, Wai SN, Kusch H, Engelmann S, Hecker M et al (2008) Proteomic characterization of the whole secretome of *Legionella pneumophila* and functional analysis of outer membrane vesicles. Infect Immun 76:1825–1836

Henry T, Pommier S, Journet L, Bernadac A, Gorvel JP, Lloubes R (2004) Improved methods for producing outer membrane vesicles in gram-negative bacteria. Res Microbiol 155:437–446

Hoekstra D, van der Laan JW, de Leij L, Witholt B (1976) Release of outer membrane fragments from normally growing *Escherichia coli*. Biochim Biophys Acta 455:889–899

Horstman AL, Kuehn MJ (2000) Enterotoxigenic *Escherichia coli* secretes active heat-labile enterotoxin via outer membrane vesicles. J Biol Chem 275:12489–12496

Hozbor D, Rodriguez ME, Fernandez J, Lagares A, Guiso N, Yantorno O (1999) Release of outer membrane vesicles from *Bordetella pertussis*. Curr Microbiol 38:273–278

Hynes SO, Keenan JI, Ferris JA, Annuk H, Moran AP (2005) Lewis epitopes on outer membrane vesicles of relevance to *Helicobacter pylori* pathogenesis. Helicobacter 10:146–156

Iwami J, Murakami Y, Nagano K, Nakamura H, Yoshimura F (2007) Further evidence that major outer membrane proteins homologous to OmpA in *Porphyromonas gingivalis* stabilize bacterial cells. Oral Microbiol Immunol 22:356–360

Kadurugamuwa JL, Beveridge TJ (1995) Virulence factors are released from *Pseudomonas aeruginosa* in association with membrane vesicles during normal growth and exposure to gentamicin: a novel mechanism of enzyme secretion. J Bacteriol 177:3998–4008

Kato S, Kowashi Y, Demuth DR (2002) Outer membrane-like vesicles secreted by *Actinobacillus actinomycetemcomitans* are enriched in leukotoxin. Microb Pathog 32:1–13

Knox KW, Vesk M, Work E (1966) Relation between excreted lipopolysaccharide complexes and surface structures of a lysine-limited culture of *Escherichia coli*. J Bacteriol 92:1206–1217

Kulp A, Kuehn MJ (2010) Biological functions and biogenesis of secreted bacterial outer membrane vesicles. Annu Rev Microbiol 64:163–184

Lazzaroni JC, Germon P, Ray MC, Vianney A (1999) The Tol proteins of *Escherichia coli* and their involvement in the uptake of biomolecules and outer membrane stability. FEMS Microbiol Lett 177:191–197

Lee EY, Bang JY, Park GW, Choi DS, Kang JS et al (2007) Global proteomic profiling of native outer membrane vesicles derived from *Escherichia coli*. Proteomics 7:3143–3153

Lee EY, Choi DS, Kim KP, Gho YS (2008) Proteomics in gram-negative bacterial outer membrane vesicles. Mass Spectrom Rev 27:535–555

Llamas MA, Ramos JL, Rodriguez-Herva JJ (2000) Mutations in each of the tol genes of *Pseudomonas putida* reveal that they are critical for maintenance of outer membrane stability. J Bacteriol 182:4764–4772

Loeb MR (1974) Bacteriophage T4-mediated release of envelope components from *Escherichia coli*. J Virol 13:631–641

Loeb MR, Kilner J (1978) Release of a special fraction of the outer membrane from both growing and phage T4-infected *Escherichia coli* B. Biochim Biophys Acta 514:117–127

Lommatzsch J, Templin MF, Kraft AR, Vollmer W, Holtje JV (1997) Outer membrane localization of murein hydrolases: MltA, a third lipoprotein lytic transglycosylase in *Escherichia coli*. J Bacteriol 179:5465–5470

Mashburn-Warren L (2008) Quinolone trafficking via outer membrane vesicles in *Pseudomonas aeruginosa*. Dissertation for the Doctor of Philosophy Thesis, University of Texas, Austin

McBroom AJ, Kuehn MJ (2005) Outer membrane vesicles In: III RC (ed) EcoSal—Escherichia coli and Salmonella : Cellular and Molecular Biology, American Society for Microbiology Press, Washington

McBroom AJ, Kuehn MJ (2007) Release of outer membrane vesicles by gram-negative bacteria is a novel envelope stress response. Mol Microbiol 63:545–558

McBroom AJ, Johnson AP, Vemulapalli S, Kuehn MJ (2006) Outer membrane vesicle production by *Escherichia coli* is independent of membrane instability. J Bacteriol 188:5385–5392

Misra RV, Horler RS, Reindl W, Goryanin II, Thomas GH (2005) EchoBASE: an integrated post-genomic database for *Escherichia coli*. Nucleic Acids Res 33:D329–D333

Mug-Opstelten D, Witholt B (1978) Preferential release of new outer membrane fragments by exponentially growing *Escherichia coli*. Biochim Biophys Acta 508:287–295

Nguyen TT, Saxena A, Beveridge TJ (2003) Effect of surface lipopolysaccharide on the nature of membrane vesicles liberated from the gram-negative bacterium *Pseudomonas aeruginosa*. J Electron Microsc (Tokyo) 52:465–469

Post DM, Zhang D, Eastvold JS, Teghanemt A, Gibson BW, Weiss JP (2005) Biochemical and functional characterization of membrane blebs purified from *Neisseria meningitidis* serogroup B. J Biol Chem 280:38383–38394

Rolhion N, Barnich N, Claret L, Darfeuille-Michaud A (2005) Strong decrease in invasive ability and outer membrane vesicle release in Crohn's disease-associated adherent-invasive *Escherichia coli* strain LF82 with the yfgL gene deleted. J Bacteriol 187:2286–2296

Rothfield L, Pearlman-Kothencz M (1969) Synthesis and assembly of bacterial membrane components. A lipopolysaccharide-phospholipid-protein complex excreted by living bacteria. J Mol Biol 44:477–492

Sabra W, Lunsdorf H, Zeng AP (2003) Alterations in the formation of lipopolysaccharide and membrane vesicles on the surface of *Pseudomonas aeruginosa* PAO1 under oxygen stress conditions. Microbiology 149:2789–2795

Schooling SR, Beveridge TJ (2006) Membrane vesicles: an overlooked component of the matrices of biofilms. J Bacteriol 188:5945–5957

Shoberg RJ, Thomas DD (1995) *Borrelia burgdorferi* vesicle production occurs via a mechanism independent of immunoglobulin M involvement. Infect Immun 63:4857–4861

Smalley JW, Birss AJ, McKee AS, Marsh PD (1991) Haemin-restriction influences haemin-binding, haemagglutination and protease activity of cells and extracellular membrane vesicles of *Porphyromonas gingivalis* W50. FEMS Microbiol Lett 69:63–67

Song T, Mika F, Lindmark B, Liu Z, Schild S et al (2008) A new *Vibrio cholerae* sRNA modulates colonization and affects release of outer membrane vesicles. Mol Microbiol 70:100–111

Sonntag I, Schwarz H, Hirota Y, Henning U (1978) Cell envelope and shape of *Escherichia coli*: multiple mutants missing the outer membrane lipoprotein and other major outer membrane proteins. J Bacteriol 136:280–285

Stephens DS, Edwards KM, Morris F, McGee ZA (1982) Pili and outer membrane appendages on *Neisseria meningitidis* in the cerebrospinal fluid of an infant. J Infect Dis 146:568

Suzuki H, Nishimura Y, Yasuda S, Nishimura A, Yamada M, Hirota Y (1978) Murein-lipoprotein of *Escherichia coli*: a protein involved in the stabilization of bacterial cell envelope. Mol Gen Genet 167:1–9

Tan TT, Morgelin M, Forsgren A, Riesbeck K (2007) *Hemophilus influenzae* survival during complement-mediated attacks is promoted by *Moraxella catarrhalis* outer membrane vesicles. J Infect Dis 195:1661–1670

Tetz VV, Rybalchenko OV, Savkova GA (1990) Ultrastructural features of microbial colony organization. J Basic Microbiol 30:597–607

Wai SN, Takade A, Amako K (1995) The release of outer membrane vesicles from the strains of enterotoxigenic *Escherichia coli*. Microbiol Immunol 39:451–456

Wensink J, Witholt B (1981) Outer-membrane vesicles released by normally growing *Escherichia coli* contain very little lipoprotein. Eur J Biochem 116:331–335

Work E, Knox KW, Vesk M (1966) The chemistry and electron microscopy of an extracellular lipopolysaccharide from *Escherichia coli*. Ann NY Acad Sci 133:438–449

Yem DW, Wu HC (1978) Physiological characterization of an *Escherichia coli* mutant altered in the structure of murein lipoprotein. J Bacteriol 133:1419–1426

Yonezawa H, Osaki T, Kurata S, Fukuda M, Kawakami H et al (2009) Outer membrane vesicles of *Helicobacter pylori* TK1402 are involved in biofilm formation. BMC Microbiol 9:197

Zhou L, Srisatjaluk R, Justus DE, Doyle RJ (1998) On the origin of membrane vesicles in gram-negative bacteria. FEMS Microbiol Lett 163:223–228

Chapter 8
Release of Outer Membrane Vesicles: Is it a Novel Secretion Mechanism?

Abstract From a comparison of the role of membranes in eukaryotic cells in the secretion of viruses and other materials it was originally proposed in 1967 that the formation and release of outer membrane vesicles (OMVs) from a Gram-negative bacterial cell represented a novel secretion mechanism of these bacteria. Subsequent experimental studies including genetic analysis supported and established the original idea. While discussing these aspects it is now proposed that the formation and release of OMVs represented the seventh secretion mechanism of Gram-negative bacteria in general.

Keywords OMVs · Secretion mechanism · Eukaryotic cells · Prokaryotes

The mechanism of production and release of OMVs by *Vibrio cholerae* cells was compared with the role of membranes in the secretory mechanisms of several eukaryotic cells (Chatterjee and Das 1967). The secretory mechanism of eukaryotic cells as recorded by microanatomical studies involved considerable activity of the cell membrane (Bennett 1956). Substances that cannot diffuse through the membrane may be incorporated into cells by a mechanism known as pinocytosis (Palade 1953). Although several mechanisms for the secretion of cellular products may exist, the mechanism of reversed pinocytosis (Palade 1961) has been observed in many cells. In this process, the secreted substances are in some way enclosed in an intracellular vesicle, which then migrates towards, adheres to, and finally communicates or fuses with the cell membrane. In addition, the extrusion of so-called colloid bodies from the thyroid cell to the vesicle occurs by the budding-off of apical cytoplasm (Lever 1961). A similar mechanism was observed in the secretory process of the rabbit apocrine sweat glands (Kurosumi 1962). A similar mechanism was also noted during the escape of certain viruses from cells (Epstein 1962). In the background of this information about the role of bounding membrane in the secretory process of certain eukaryotic cells, the budding-off of the vesicles from the surface of actively growing *V. cholerae* cells was interpreted to represent

S. N. Chatterjee and K. Chaudhuri, *Outer Membrane Vesicles of Bacteria*,
SpringerBriefs in Microbiology, DOI: 10.1007/978-3-642-30526-9_8,
© The Author(s) 2012

a secretory mechanism of the bacterial cells (Chatterjee and Das 1966, 1967). It was argued that the non diffusible enzymes and other chemicals of the vibrios normally located in the periplasm were secreted through the OMVs (Chatterjee and Das 1967). Subsequently, different secretion mechanisms operating in Gram-negative bacteria were discovered and worked out in detail.

Secretion by any cell is a genetically controlled active process requiring expenditure or transduction of energy. Gram-negative bacteria have evolved six major secretory pathways depending on the Sec or Tat signal sequence (Lee et al. 2008). The Sec-dependent secretory pathways utilize a cleavable N-terminal signal peptide for protein transport across the inner membrane (Kostakioti et al. 2005) and include the type II (T2SS), IV (T4SS), and V (T5SS) secretion processes. Of these, the T2SS pathway utilizes both the Sec and Tat signal sequences, is known as the general secretory pathway, and has found use in the secretion of several toxins (Voulhoux et al. 2001). The transfer of DNA and some multisubunit toxins, including pertussis toxin, is governed by the T4SS secretion mechanism (Cascales and Christie 2003). Both Sec-dependent and Sec-independent secretion processes were observed in the T4SS mechanism (Desvaux et al. 2004). In the T5SS system, proteins are translocated across the outer membrane via a trans-membrane pore formed by a self-encoded ß-barrel structure (Desvaux et al. 2004). This T5SS system is also known as the autotransporter pathway. The T1SS, T3SS, and T6SS systems do not involve periplasmic intermediates and are one-step mechanisms. Various molecules from ions and drugs to proteins are secreted by the T1SS system via an ATP binding cassette transporter like channel (Binet and Wandersman 1995). Some pathogenic bacteria utilize the T3SS system that allows direct injection of a protein into a eukaryotic host cell (Galan and Collmer 1999). The T6SS pathway has been reported to be utilized by *Pseudomonas aeruginosa* and *V. cholerae* for secretion of several proteins (Mougous et al. 2006; Pukatzki et al. 2006), although the cholera toxin CT is known to be secreted by the T2SS pathway.

In the background of this knowledge of the different secretion mechanisms displayed by Gram-negative bacteria, the release of OMVs by the actively growing cells represents a totally different process. It was originally believed to be a novel secretion mechanism (Chatterjee and Das 1966, 1967). It is not a passive mech-anism of release of different cargo through the OMVs as it involves a definite sorting process. Some components of bacterial cells are enriched and others are excluded from the OMVs vis-à-vis their concentration in the whole bacterium. McBroom and Kuehn (2007) also drew analogy with the membrane-involving secretion processes in eukaryotic cells involving membrane-bound vesicles and believed that the capability extends to prokaryotes as well. The different purposes of the secretion through membrane-bound micro particles of the eukaryotic cells include specific elimination of undesirable compounds from the cell (Pilzer et al. 2005). In comparison, the specific release of envelope material via OMVs extends this paradigm, the authors believed, to Gram-negative bacteria and fulfills the essential requirements of a genuine prokaryotic secretion process (McBroom and Kuehn 2007). Vesicle formation by Gram-negative bacteria thus forms a specific secretory mechanism, one that is clearly different from any of the six different mechanisms known thus far (Kuehn and Kesty 2005). A genetic study of several vesiculation mutants of *Escherichia coli* revealed that vesiculation was not a consequence of bacterial lysis, could not be correlated with membrane instability, and may be considered a fundamental characteristic of Gram-negative bacterial growth (McBroom et al. 2006). It was further demonstrated that vesiculation fulfilled the key requirements of a genuine prokaryotic secretion process (McBroom and Kuehn 2007).

Following the elegant review by Kulp and Kuehn it is now clear that the release of OMVs by Gram-negative bacteria possesses a number of distinctive features to distinguish it from their other six secretion mechanisms (Kulp and Kuehn 2010): (1) bacterial lipids, membrane proteins, and other hydrophobic molecules can be secreted from the parent cells; (2) materials released or secreted via the OMVs are protected from the action of enzymes or other chemicals present in the surrounding milieu; (3) OMVs allow the delivery of a high concentration of the cargo to distal sites without being affected or degraded by enzymes, nucleases, and the like present, if any, in the surrounding milieu; (4) ligands bound to the surface of OMVs can bind to the receptors on the surface of host or target cells; and (5) naked molecules that cannot have entry into the host or target cells may enter the cells when transferred via the lumen of the OMVs. The outer membrane and periplasmic proteins, in particular, are packaged into the OMVs under different growth conditions of the bacteria, indicating that the formation and release of OMVs is an energy sink (Kuehn and Kesty 2005). Up-to-date evidence shows that many different cargos (proteins, lipids, genetic materials, etc.) are encapsulated and released via the OMVs. Export of cytolysin A (ClyA) into the extracellular medium is not governed by any of the six secretion mechanisms discussed earlier but is mediated via OMVs (Wai et al. 2003). Therefore, following the original proposal of Chatterjee and Das (1966, 1967), it has now been established, as shown by Kuehn and Kesty (2005), that the release of OMVs represents a novel secretion pathway of Gram-negative bacteria. Accordingly, it is now proposed that the mechanism of release of proteins and other materials through OMVs be henceforth referred to as the type VII (T7SS) secretion pathway of Gram-negative bacteria. However, the genetics of this secretion pathway and the mechanism of energy transfer involved remain to be worked out in detail.

References

Bennett HS (1956) The concepts of membrane flow and membrane vesiculation as mechanisms for active transport and ion pumping. J Biophys Biochem Cytol 2:99–103

Binet R, Wandersman C (1995) Protein secretion by hybrid bacterial ABC-transporters: specific functions of the membrane ATPase and the membrane fusion protein. EMBO J 14:2298–2306

Cascales E, Christie PJ (2003) The versatile bacterial type IV secretion systems. Nat Rev Microbiol 1:137–149

Chatterjee SN, Das J (1966) Secretory activity of *Vibrio cholerae* as evidenced by electron microscopy. In: Uyeda R (ed) Electron microscopy. Maruzen Co. Ltd, Tokyo

Chatterjee SN, Das J (1967) Electron microscopic observations on the excretion of cell-wall material by *Vibrio cholerae*. J Gen Microbiol 49:1–11

Desvaux M, Parham NJ, Henderson IR (2004) The autotransporter secretion system. Res Microbiol 155:53–60

Epstein MA (1962) Functional aspects of the structure of some animal viruses. Br Med Bull 18:183–186

Galan JE, Collmer A (1999) Type III secretion machines: bacterial devices for protein delivery into host cells. Science 284:1322–1328

Kostakioti M, Newman CL, Thanassi DG, Stathopoulos C (2005) Mechanisms of protein export across the bacterial outer membrane. J Bacteriol 187:4306–4314

Kuehn MJ, Kesty NC (2005) Bacterial outer membrane vesicles and the host-pathogen interaction. Genes Dev 19:2645–2655

Kulp A, Kuehn MJ (2010) Biological functions and biogenesis of secreted bacterial outer membrane vesicles. Annu Rev Microbiol 64:163–184

Kurosumi K (1962) Some morphological aspects of the secretory activities of various glandular cells. In: Breese SS (ed) Electron microscopy. Academic Press, London

Lee EY, Choi DS, Kim KP, Gho YS (2008) Proteomics in Gram-negative bacterial outer membrane vesicles. Mass Spectrom Rev 27:535–555

Lever JD (1961) Electron microscopy in anatomy. Arnold, London

McBroom AJ, Kuehn MJ (2007) Release of outer membrane vesicles by Gram-negative bacteria is a novel envelope stress response. Mol Microbiol 63:545–558

McBroom AJ, Johnson AP, Vemulapalli S, Kuehn MJ (2006) Outer membrane vesicle production by *Escherichia coli* is independent of membrane instability. J Bacteriol 188:5385–5392

Mougous JD, Cuff ME, Raunser S, Shen A, Zhou M et al (2006) A virulence locus of *Pseudomonas aeruginosa* encodes a protein secretion apparatus. Science 312:1526–1530

Palade GE (1953) Fine structure of blood capillaries. J Appl Physiol 24:1424–1436

Palade GE (1961) The secretory process of the pancreatic cell. In: Boyd JD, Johnson FR, Lever JD (eds) Électron Microscopy in Anatomy. Arnold, London

Pilzer D, Gasser O, Moskovich O, Schifferli JA, Fishelson Z (2005) Emission of membrane vesicles: roles in complement resistance, immunity and cancer. Springer Semin Immunopathol 27:375–387

Pukatzki S, Ma AT, Sturtevant D, Krastins B, Sarracino D et al (2006) Identification of a conserved bacterial protein secretion system in *Vibrio cholerae* using the Dictyostelium host model system. Proc Natl Acad Sci U S A 103:1528–1533

Voulhoux R, Ball G, Ize B, Vasil ML, Lazdunski A, Wu LF, Filloux A (2001) Involvement of the twin-arginine translocation system in protein secretion via the type II pathway. EMBO J 20:6735–6741

Wai SN, Lindmark B, Soderblom T, Takade A, Westermark M et al (2003) Vesicle-mediated export and assembly of pore-forming oligomers of the enterobacterial ClyA cytotoxin. Cell 115:25–35

Chapter 9
Outer Membrane Vesicles: Physiological Medical Applications

Abstract Outer membrane vesicles (OMVs) produced by Gram-negative bacteria exhibit enormous functional diversity depending on the bacterial species and environmental niche: these have enabled various physiological roles for these vesicles to play and at the same time OMVs have found important medical applications. The interaction of OMVs with the host can lead to varied innate immune responses such as direct interaction with immune cells, triggering of cell-mediated immunity, or activation of proinflammatory response leading to cytotoxicity. OMVs were found to carry antigenically active virulence factors and the potential of OMVs for nonreplicating vaccines has been explored in several Gram-negative organisms in animal models. OMVs are being licensed and used for vaccination in combating *Neisseria meningitidis* serogroup B infections.

Keywords Innate immune response · B cell · Complement system · Cell-mediated immunity · Proinflammatory response · Immunogen · Vaccine · Detergent-treated OMVs · Native OMVs · Combination vaccines · Recombinant OMVs vaccines · Reverse vaccinology · Adjuvant

9.1 Activation of Innate Immune Response

Inside the host, the bacteria first encounter the innate immune system or the first line of host defense. The innate immune system is a response system designed to alert the host rapidly of the presence of a microbial pathogen trying to breach the integument of the host. At the same time, the defense initiates an effort to make the internal milieu of the host free from pathogen load. The innate immune system is so-named because it is not adaptable, is intrinsic to the host, and does not change over its lifetime.

The innate response consists of two distinct parts (Diacovich and Gorvel 2010). First is recognition of the pathogen by the host. The molecular patterns present in

S. N. Chatterjee and K. Chaudhuri, *Outer Membrane Vesicles of Bacteria*,
SpringerBriefs in Microbiology, DOI: 10.1007/978-3-642-30526-9_9,
© The Author(s) 2012

the pathogen (called pathogen-associated molecular patterns (PAMPs)) serve as the recognition signal of the pathogen for the host. PAMPs are either recognized by the receptor present on the host cell surface and bind to it, or the soluble proteins/serum proteins are recognized and altered by PAMPs. Once the pathogen is recognized, a signaling cascade is activated that involves the recruitment of destructive effector mechanisms for killing and eliminating the pathogen. The innate immune response employs a number of strategies depending on the host and the pathogen such as (a) an immune cell such as neutrophil recruitment to the infection site, mediated by chemical factors including specialized proinflammatory chemical mediators or messengers of the immune system known as cytokines controlling bacterial numbers until a full immune response is mounted; (b) a complement cascade activation which leads to the formation of a membrane-attack-complex that initiates bacterial cell lysis and promotes clearance of dead cells or antibody complexes, (c) identification and removal of foreign substances present in blood, lymph, tissues, and organs of the host by specialized white blood cells such as natural killer cells, mast cells, eosinophils, basophils, and phagocytic cells such as macrophages, dendritic cells, and neutrophils; and (d) adaptive immune system activation via a process known as antigen presentation.

Outer membrane vesicles (OMVs) produced by bacteria are spherical bodies bounded by bacterial outer membrane and contain LPS, outer membrane proteins, lipids, lipoproteins, flagellin, and deoxy-cytidylate-phosphate-deoxy-guanylate (CpG) DNA as well as soluble periplasmic materials including the virulence factors and enzymes. Many of these, such as LPS, flagellin, outer membrane proteins, and the like, constitute PAMPs (Beveridge 1999; Kuehn and Kesty 2005). PAMPs are unique microbial molecules that are recognized by a variety of host cell receptors called pattern recognition receptors (PRR) that bind specifically to the conserved portions of these molecules. PRRs are present on specific host cell types such as macrophages, dendritic cells, endothelial cells, mucosal epithelial cells, and lymphocytes. The molecular composition of vesicles is dependent on the bacterial strain of origin and the combination of PAMPs, adhesins, invasins, and other outer membrane components make the OMVs recognizable by the immune system. Several lines of evidence suggest that OMVs produced by colonizing pathogens have a complex and as yet unexplored impact on the immune response, leading to clearance of the pathogen or enhancement of virulence or both depending on the pathogenic organism releasing OMVs. Bacterial OMVs mount various immune responses, the most recent information in this area is discussed below.

9.2 B-Cell Activation by OMVs

B-cell activation by OMVs is exemplified by *Moraxella catarrhalis* outer membrane vesicles (OMVs) which activates B cells through a T-cell independent mechanism. *M. catarrhalis,* an exclusive human respiratory pathogen, is a

Gram-negative aerobic diplococcus. It is responsible for about 10 % of the total increase in severity of chronic obstructive pulmonary disease (COPD) and contributes significantly to chronic airway inflammation, which is a hallmark of the disease (Sethi et al. 2007). *M. catarrhalis*, like other pathogenic organisms, uses multiple strategies to evade and dampen the immune response of the host and establishes itself as a successful pathogen. This leads to colonization and eventually invasion in the host. Some of the strategies include attachment through adhesins to the nasopharyngeal epithelium, recognition of host extracellular matrix proteins by PAMPs, evasion of the host complement system, and redirection of the immune system (Vidakovics et al. 2010).

As do other Gram-negative pathogens, *M. catarrhalis* also produce OMVs in vitro and in vivo. It has been demonstrated recently that OMVs from *M. catarrhalis* can activate human B cells (Vidakovics et al. 2010) isolated from pharyngeal lymphoid tissue which is the potential reservoir for *Moraxella* (Heiniger et al. 2007). OMVs interact with tonsillar B cells via superantigen *Moraxella* immunoglobulin D (IgD)-binding protein MID expressed on OMVs. MID is a 200-kDa outer membrane protein belonging to the autotransporter family (Hallstrom et al. 2008) and has a specific affinity for the human B-cell receptor (BCR), the membrane-bound IgD. MID is referred to as a B-cell activator or B-cell superantigen as it recognizes conserved regions of IgD in a nonimmune manner. Nonimmune recognition of microbial molecules is the recognition irrespective of antigenic specificity. MID molecules form a trimeric structure required for binding to IgD. The IgD binding domain is located at 962–1200 amino acid residues (Nordstrom et al. 2002). MID binds to 198–224 amino acids in the heavy chain constant region 1 (CH1) of human IgD, strongly cross-links the BCR and activates B cells in a T-independent manner (Samuelsson et al. 2006).

Data regarding on the interaction of *M. catarrhalis* OMVs with B cells emerge from some recent studies (Singh et al. 2012; Vidakovics et al. 2010). The MID containing OMVs induce BCR cross-linking forming a BCR complex consisting of the surface immunoglobulin (Ig) molecule and the Ig-α/β heterodimer(s); each of these monomers contains the Ig-superfamily tyrosine-based activation motif or ITAM in their cytoplasmic tails. Antigens from OMVs bind to the BCR complex resulting in segregation of the antigen–receptor complex into specialized membrane region or lipid rafts followed by internalization of vesicles. Cross-linking of BCR leads to phosphorylation of ITAM motifs and recruitment of kinases such as tyrosine kinase Syk and its autophosphorylation. These protein tyrosine kinases (PTKs) facilitate a number of downstream signaling pathways such as activation of lipid-metabolizing enzymes phosphoinositide 3-kinase (PI3-K) and phospholipase C (PLC) $\gamma2$, and generation of increased intracellular Ca^{2+} influx. These events further result in the activation of multiple signaling pathways including activation of p38 mitogen-activated protein (MAP)-kinase and nuclear factor (NF)kB (Kurosaki et al. 2010). In *Morexella*, B-cell activation was mediated in a Toll-like receptor (TLR)-dependent manner; TLR2 and TLR9 were especially involved in the activation of B cells by *Morexella* PAMPs. The receptors TLR2 and TLR9 on the host cell surface also participate in the signaling induced by OMVs. TLRs are the Toll-like receptors present on the host cell surface. TLRs recognize conserved microbial patterns and upon binding of the specific ligand, they dimerize to form functional receptors. TLR2 recognizes structurally diverse PAMPs such as peptidoglycan (PG) and lipoproteins. Binding of PAMPs–TLRs initiates a signaling cascade via recruitment of myeloid differentiation 88 (MyD88), which transmits the signal through the tumor necrosis factor (TNF), receptor-associated factor 6 (TRAF 6) and transforming growth factor β-activated kinase 1 (TAK-1) to activate NFkB and MAPK pathways (Kawai and Akira 2010). OMVs are internalized in endosomes (Vidakovics et al. 2010) which have a preponderance of TLR9. TLR9 recognizes nucleic acid as a ligand, mainly single-stranded DNA containing CpG dinucleotide motif in which the cytosine is unmethylated (Chaturvedi and Pierce 2009). Thus binding of superantigen MID–BCR initiates the signaling event leading to overlapping activation of the NFkB and MAPK pathways, which, in turn, promotes enhanced activation of transcription factors entailing proinflammatory cytokine such as interleukin 6 (IL-6) secretion and B-cell activation. Additionally, B cells produce non-specific IgM, which are not directed against *M. catarrhalis*. The B-cell activation pathway is illustrated in Fig. 9.1. Vesicle secretion provides *Moraxella* with a sophisticated mechanism to modify the host immune response, avoiding contact between bacteria and the host.

Similar to *M. catarrhalis*, OMVs produced by *Neisseria lactamica* induce B-cell proliferative response leading to the expression of polyclonal IgM. This proliferative B-cell response is highly conserved among divergent strains of *N. lactamica*. It is restricted to the B-cell pool including B cells containing a cluster of differentiation antigen 5 (CD5$^+$ B cells) and is independent of T-cell interaction. This mitogenic B-cell response is mediated by BCR. B-cell superantigens on *N. lactamica* OMVs bind to IgD and IgM on B cells and stimulate activation. Treatment of tonsillar mucosal mononuclear cells (TNMCs) with the pronase-reducing surface Ig or anti-BCR antibody resulted in abrogation of proliferative response, suggesting the involvement of BCR. The molecular component on *N. lactamica* OMV responsible for the B-cell response could be a protein. Lipo-oligosaccharide (LOS) or PorB proteosome from *N. lactamica* failed to proliferate B cells. In another study on mouse B cells and *Nesisseria* OMVs, it was shown in vitro that the B-cell mitogenic effect of OMVs is not critically dependent on the presence of LPS suggesting that it was not mediated by theTLR4 machinery (Durand et al. 2009). On the other hand, loss of proliferation was observed after protease digestion of OMVs (Vaughan et al. 2010). Because *N. lactamica* OMVs express more than 180 proteins, the protein component is yet to be determined.

N. lactamica OMV harbors a B-cell superantigen with high affinity for IgM and IgD that induces the activation of innate B-cell subsets including CD5$^+$ B1 cells and marginal zone cells, resulting in the production of natural antibodies (Abs) which is critical for maintaining immunological ignorance in the host (Vaughan et al. 2010). No contribution of TLR9 was observed during activation of B cells by *N. lactamica* OMVs.

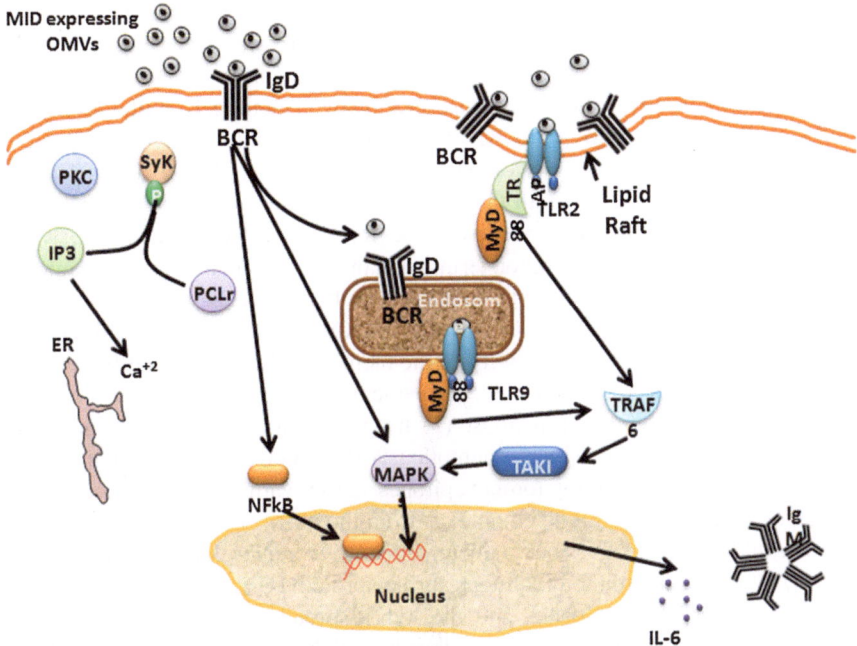

Fig. 9.1 B-cell activation by *M. catarrhalis* OMVs. *M.catarrhalis* OMVs express IgD-binding protein MID which interacts with the surface IgD receptor on B-cells. OMVs are internalized by interaction of lipoproteins with TLR2 leading to lipid raft formation. TLR9 further recognizes OMV-associated DNA and TLR9 together with TLR2 activates a signaling cascade involving MyD88, TRAF6, TAK1, p38 MAP kinase, and NF-kB finally leading to il-6 secretion. Additionally, B cells produce IgM which are nonspecific and not directed against *Morexella*

9.3 Interaction of OMVs with Complement System

In *Morexella*, OMVs have been demonstrated to interact with the alternative pathway of the complement system (Tan et al. 2007). The complement system is a key component of the innate immune response and plays a central role the in defense against invading bacteria. It is involved in opsonization and clearing the pathogens from the host and complements the function of Abs and phagocytic cells. The complement system consists of a number of proteins circulating in the blood and bathing the fluids surrounding tissues. The proteins normally circulate as inactive precursors, but are stimulated in response to the molecular components of the microorganism. They initiate a cascade wherein the binding of one protein promotes the binding of the next protein in the cascade. Three major biochemical pathways have been recognized for activation of the complement system: the classical complement pathway, the lectin pathway, and the alternative complement pathway (Hallstrom and Riesbeck 2010; Singh et al. 2010) as illustrated in Fig. 9.2. The classical complement pathway is activated by bacterial LPS, nucleic acids, and antigen–antibody complexes. The lectin-mediated pathway is activated by the interaction of mannose residues on the bacterial cell surface with mannose-binding lectin in the plasma and tissue fluids, and the alternative pathway is activated by binding of C3b protein directly to microbial surfaces and to antibody molecules. Pathogen recognition is followed by opsonization of the pathogen or alteration of the pathogen by opsonins so as to be efficiently engulfed by phagocytes leading subsequently to phagocytosis (Blom et al. 2009; Zipfel and Skerka 2009). The end result and defense function of all three pathways are the same. All three pathways lead to cleavage of the component C3 and formation of C5 convertase, which subsequently activates the common terminal pathway. This pathway is so-named as it leads to the end-product, that is the "membrane attack complex" (MAC) containing the components C5b, C6, C7, C8, and C9 and facilitates the killing of the pathogen by altering the permeability of the membrane leading to cell lysis (Singh et al. 2010). Proteins produced by all complement pathways are involved in carrying out the following defense functions: (1) triggering inflammation, (2) attracting phagocytes to the infection site by chemotaxis, (3) promoting attachment of antigen, (4) lysing pathogen displaying foreign epitopes, (5) activating na B lymphocytes and (6) removing harmful immune complexes from the host.

Morexella OMVs have been documented to contain the virulence factors UspA1/A2 belonging to the ubiquitous surface protein (Usp) family and consist of three variants: UspA1, UspA2, and the hybrid protein designated as UspA2H. UspA1/A2 are oligomeric coiled-coil adhesins with binding sites for host epithelial cell-associated macromolecules such as fibronectin and basement membrane-associated laminin and cell adhesion protein vitronectin (Attia et al. 2006; Tan et al. 2005). Recently, it has been shown that UspA1/A2 is involved in the serum resistance of *M. catarrhalis* by binding noncovalently to the third component of the complement system (C3) which is inactivated. In addition, UspA1/A2 interacts with vitronectin and C4b-binding protein (C4bBP). Thus, UspA1/A2 is the main virulence factor involved in the complement resistance of *M. catarrhalis*. UspA1/A2 binds and inactivates complement factor C3 in a dose-dependent manner and restrains complement activation. OMVs from wild-type clinical strains of *M. catarrhalis* bound to C3 and counteracted the complement cascade to a larger extent than did OMVs deficient in UspA1/A2 (Tan et al. 2007). In an earlier study it was shown that only *Morexella* OMVs containing UspA1/A2 are capable of protecting the respiratory pathogen *H. influenzae* from complement-mediated killing by inactivation of C3. These results suggested a novel OMV-mediated collaborative strategy employed by pathogenic bacteria to defeat innate immunity.

In contrast, native and detergent-treated *Neisseria meningitidis* OMVs exhibited substantial complement activation in vitro in whole blood anticoagulated with the recombinant hirudin analogue lepirudin. The key activation products were measured following incubation with *N. meningitidis* native OMVs, detergent-treated OMVs, LPS, and wild-type bacteria. Compared to OMVs, *N. meningitidis* LPS was a weak complement activator whereas OMVs had comparable activation with wild-type bacteria (Bjerre et al. 2002).

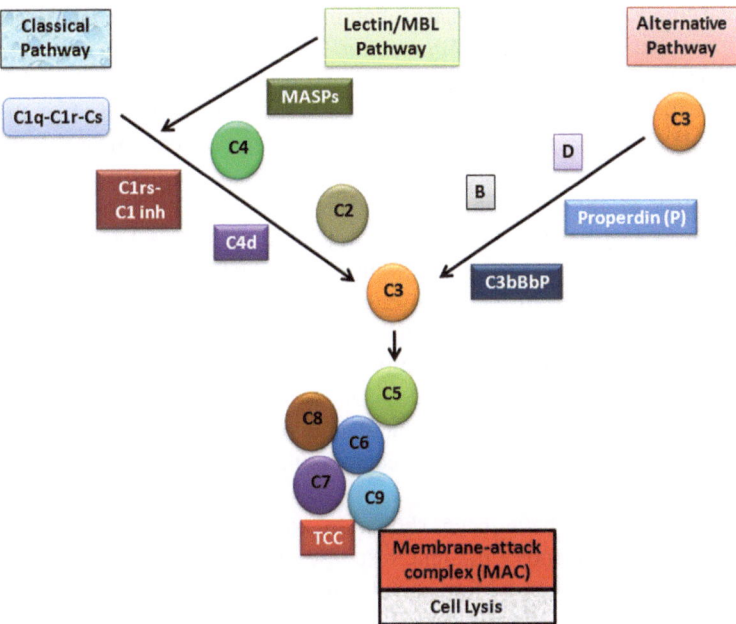

Fig. 9.2 The three pathways of activation of complement cascade during bacterial infection

9.4 OMV Triggers Cell-Mediated Immunity

Salmonella OMVs provide an example for triggering cell-mediated immunity in the host. *Salmonella typhimurium* replicate in the intracellular vacuoles of phagocytes resulting in a systemic typhoidlike disease (Jones and Falkow 1996; Lowrie et al. 1979). *Salmonella* outer membranes express classical inflammatory molecules such as lipopolysaccharide (LPS) and lipoproteins but modify the bacterial envelope through processes controlled by the PhoP/PhoQ regulatory system and resist innate immune recognition (Ernst et al. 2001). CD4[+] T-cell responses are an essential component of immunity to *Salmonella* (Mastroeni et al. 1993), and FliC, the major subunit protein of bacterial flagella, is an antigen recognized by CD4[+] T cells via TLR5, from both *Salmonella*-infected mice and humans (Cookson and Bevan 1997; McSorley et al. 2000; Sztein et al. 1994).

It has been demonstrated that *Salmonella* produces OMVs containing OmpA and ligands that can stimulate TLRs and also contain the natural antigens protected from protease digestion and recognized by murine CD4[+] T cells responding to *Salmonella* infection. Thus OMV properties are linked with innate and adaptive host defense mechanisms (Bergman et al. 2005).

Another later study demonstrated that OMVs from *Salmonella typhimurium* stimulated innate proinflammatory responses from professional antigen presenting cells (APCs) such as macrophages and dendritic cells in vitro (Alaniz et al. 2007). During *Salmonella* infection, macrophages are critical phagocytes that recognize microbial components, initiate innate inflammatory response, and clear *Salmonella*. It has been demonstrated that *Salmonella* OMVs are structures recognized by macrophages which, in turn, are activated and produce two important proinflammatory mediators, TNF-α and nitric oxide (NO). Dendritic cells (DCs), on the other hand, are APCs that express surface molecules and produce distinct inflammatory signals which dictate the nature of the adaptive immune response. DCs are functional entities at the junction of the innate and adaptive immune responses. *Salmonella* OMVs have been shown to efficiently stimulate DCs to mature, as evidenced by increased expression of surface major histocompatibility complex class II (MHC-II) and cluster of differentiation 86 (CD86), and produce proinflammatory cytokines, TNF alpha (TNF-α) and interleukin-12 (IL-12). The induction pathway of cell-mediated immunity by OMVs is illustrated schematically in Fig. 9.3. Furthermore, OMV-mediated DC activation takes place in a TLR4-dependent as well as TLR4-independent manner. Because TLR4 recognizes LPS, it is consistent with the fact that OMVs contain other distinct PAMPs in addition to LPS.

Salmonella OMVs have been shown to contribute to the development of *Salmonella*-specific adaptive immune response in vivo, or in other words, OMVs possess protective antigens. More specifically, it has been demonstrated that (1) *Salmonella* OMVs possess B- and T-cell antigens recognized by *Salmonella* immune mice, and (2) OMVs' prime *Salmonella*-specific immunity in vivo, limits bacterial replication in vivo and protects mice from infection (Alaniz et al. 2007). Also mice injected intraperitoneally with *Salmonella* OMVs stimulated proinflammatory response from professional APCs and prime to *Salmonella*-specific adaptive responses in vivo (Alaniz et al. 2007; Bergman et al. 2005).

9.5 OMV Stimulates Proinflammatory Response

OMVs produced by bacteria are often a key factor in initiating an inflammatory response in host epithelial cells and macrophages. *Salmonella* OMVs elicit proinflammatory responses in macrophages which, in turn, results in initiating an adaptive immune response as discussed above. In response to OMVs from *Helicobacter pylori* or *Pseudomonas aeruginosa*, epithelial cells produce proinflammatory cytokine IL-8, which is a potent activator for neutrophil and monocytes in vivo (Bauman and Kuehn 2009; Ismail et al. 2003).

H. pylori adheres to the gastric epithelial cells via multiple surface components and this adherence is a contributory factor in *H. pylori*-associated disease (Guruge et al. 1998). However, evidence suggests that gastritis and epithelial cell damage are initiated by the release of bacterial virulence factors induced in the absence of bacterial attachment to the epithelium. It is hypothesized that OMVs containing proteins, LPS, lipoproteins, and VacA modulate epithelial cell function independent of *H. pylori* adherence to gastric epithelial cells (Ismail et al. 2003). A dose-dependent cell proliferation, vacuolation, loss of cell viability, and production of

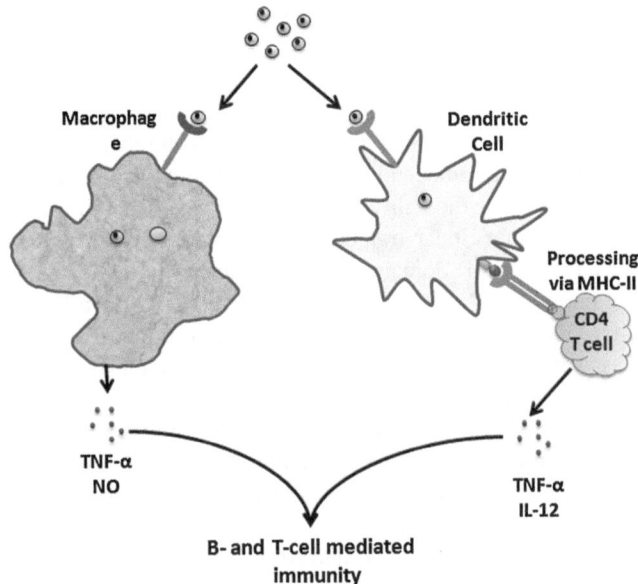

Fig. 9.3 Stimulation of cell-mediated immunity by *Salmonella* OMVs. *Salmonella* outer membrane vesicles can be taken up by macrophages or dendritic cells (*DCs*) through receptor-mediated interaction. Stimulation of macrophages leads to the production of TNF-α and nitric oxide (*NO*). Stimulation of dendritic cells leads to mature DCs, as evidenced by increased expression of MHC-II and TNF-α and IL-12. Finally activated macrophages and DCs trigger B-and T-cell mediated immune response

proinflammatory cytokines IL-8 were observed following incubation of gastric epithelial cells with *H. pylori* OMVs (Ismail et al. 2003).

Recent studies by Kaparakis and colleagues (2010) show that *H. pylori* OMVs induce innate immune responses via cytosolic host protein nucleotide binding oligomerization domain 1 (NOD1)-dependent but TLR-independent mechanism in vitro and in mice. Such responses were not observed in NOD1 knockout cells. NOD1 acts as an intracellular sensor of bacterial pathogens through its recognition of cell wall PG. Human NOD1 exhibits exquisite specificity for diaminopimelate containing GlcNac-MurNac tripeptide motif which is almost exclusively present in the PG of Gram-negative bacteria (Magalhaes et al. 2005). An intracellular presentation of bacterial PG is required to trigger NOD1 signaling in epithelial cells. OMVs are proposed to be a novel mechanism by which PG can be transported into the cytoplasm of nonphagocytic epithelial cells and initiate NOD1-dependent immune responses in vivo (Kaparakis et al. 2010). OMVs were prepared from *lysA* negative *H. pylori* strain (251*lysA*) in which tritiated meso-diaminopimelic substrate is specifically incorporated into the GM-TriDAP motif of Gram-negative PG (Viala et al. 2004). Silver deposits, corresponding to tritiated PGs, were associated with the human gastric adenocarcinoma epithelial cell line AGS that had been co-cultured with OMVs from *H. pylori* 251*lysA* cells. These

OMVs were estimated to contain approximately 0.3–0.5 μg of the muramic acid moiety per mg of OMV protein (Kaparakis et al. 2010).

NOD1 recognizes OMV-associated PGs, and lipoproteins, DNA, and LPS are not the prime agonists for NF-kB activation in epithelial cells stimulated by *H. pylori* OMVs. In addition, introduction of PGs, or GlaNac-MurNAc tripeptides into the cellular cytoplasm via microinjection did not result in NOD1-dependent signaling. Thus, OMVs might provide the right conformation of the PG to be presented to NOD1 to stimulate further intracellular trafficking. Moreover, MyD88 and Mal (MyD88 adapter such as TIRAP; TIR domain containing adapter protein) knockout mice, in which all the known TLR pathways are defective, have unvaried inflammatory responses to OMVs. In addition, pharmacological disruption of lipid rafts prevents OMV entry as well as induction of innate immune signaling in host cells. Thus, *H. pylori* OMVs enter host epithelial cells via lipid rafts to transport their PG–NOD1, which results in activation of nuclear factor kappa B(NF-kB) and inflammatory responses, such as production of IL-8 (Kaparakis et al. 2010). The NOD1-mediated inflammatory response induced by OMVs is schematically depicted in Fig. 9.4.

In addition to *H. pylori*, OMVs from *P. aeruginosa* and *Neisseria gonorrhea* were also shown to upregulate NF-κB and NOD1-dependent responses (Kaparakis et al. 2010). Collectively, OMVs from *H. pylori*, *P. aeruginosa*, and *N. gonorrhea* are able to penetrate epithelial cells through lipid rafts and elicit innate immune responses via NOD1-dependent and TLR-independent machinery. OMVs are thus proposed to be part of a widespread mechanism that allows Gram-negative bacteria to transport PGs to cytosolic NOD1, which may promote host inflammatory responses. In addition, native OMVs from *N. meningitidis* induce release of proinflammatory cytokines in vitro in a dose-dependent manner (Bjerre et al. 2002). However, a delicate balance between immunity and inflammation will dictate whether microbes will be benefited or perished.

9.6 OMVs as Immunogens

OMVs are naturally produced by a wide variety of Gram-negative pathogenic organisms during growth and are released in the environment. Such vesicles are known to contain LPS, lipoproteins, outer membrane proteins, periplasmic contents, DNA, RNA, and virulence factors. As such, the OMVs can be regarded as representing bacterial surfaces and an acellular source of bacterial antigens (Ellis and Kuehn 2010). Soon after the discovery that OMVs could be a source of active antigen, a major interest of immunotherapeutic research has been based on the examination of the potential of OMVs as immunogens. Or in other words, studies are being directed to examine the immunogenicity and protective response of OMVs using animal models. The number of studies is growing day by day and adaptive memory immune response of OMVs from several Gram-negative bacteria studied to date are discussed below. Table 9.1 includes the summary of some results showing the vaccine potential of OMVs.

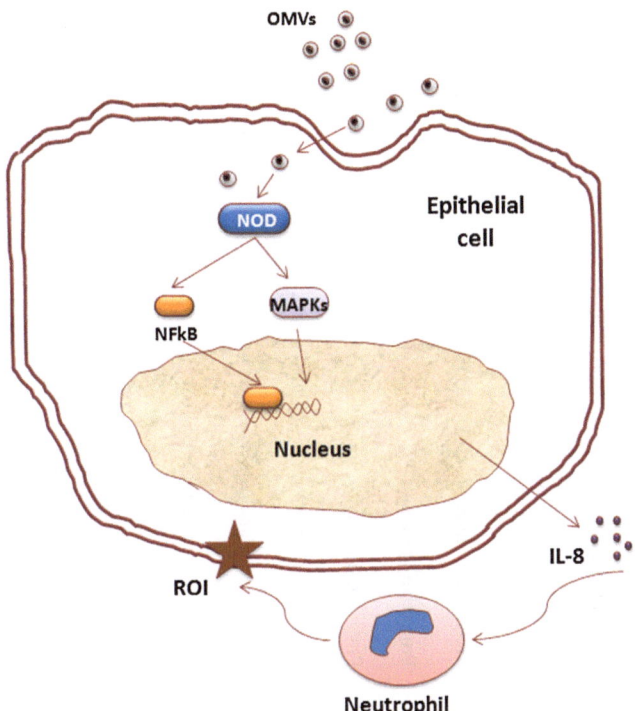

Fig. 9.4 Proinflammatory response in epithelial cells by OMVs. OMVs are internalized through lipid raft and stimulate NOD1 mediated activation of NF-kB and p38 MAP kinase. This leads to the production and secretion of proinflammatory cytokines and recruitment of neutrophils. Stimulated and infiltering neutrophils release reactive oxygen intermediates that can cause damage to the epithelium. Proinflammatory response has been demonstrated in *H. pylori*, *M. catarrhalis*, *P. aeruginosa*, and *N. meningitidis*

9.6.1 Neisseria meningitidis

Neisseria meningitidis, a Gram-negative encapsulated bacterium, is the causative agent of bacterial meningitidis in human. Capsular polysaccharide (CPS)-based vaccines are available and used for the diseases causing serogroups A, C, Y, and W135. However, these polysaccharide-based vaccines were not useful for the disease causing serogroup B of *N. meningitidis* and this serogroup accounts for about 50 % of cases in the United States (Sharip et al. 2006). The group B capsule is structurally similar to antigens expressed by neural tissues and is thus a poor immunogen and has the potential to elicit autoantibodies. Thus polysaccharide protein conjugate vaccine also may not be feasible for prevention of disease caused by group B meningococcus (Zimmer and Stephens 2006). In view of the difficulties encountered with the polysaccharide vaccines, other strategies have been sought and initial attempts mainly relied on the epidemiological data available. The age-dependent incidence of serogroup B meningococcal disease could be

Table 9.1 Evaluation of vaccine potential of outer membrane vesicles from pathogenic bacteria

Organism	OMVs	Adjuvant	Immunization and Route	Immunity	Major antigen	References
Acinetobacter baumannii	Wild type	Aluminum phosphate	Mouse; i.m	Protective immunity; IgG, IgM	n.d	(McConnell et al. 2011)
Borrelia burgdorferi	Wild type	Nil	Rabbit; i.d.	Neutralizing antibodies	OspA, DbpA, OspC	(Shang et al. 2000)
Bordetella pertussis	Detoxified	Aluminum hydroxide nil-for i.n.	Mouse; i.n., i.p.	Protective immunity	n.d	(Roberts et al. 2008)
Bordetella pertussis	PagL expressed and detoxified LOS	Nil	Mouse; i.n.	Protective immunity	n.d	(Asensio et al. 2011)
Burkholderia pseudomallei	Wild type	Nil	Mouse; s.c.	Protective immunity	n.d.	(Nieves et al. 2011)
Edwardsiella tarda	Wild type	Nil	Fish; i.p	Protective immunity, IgG	n.d	(Park et al. 2011)
Escherichia coli (ETEC)	Wild type	Nil	Mouse; i.n.	Protective immunity	EtpA, CexE, LT	(Roy et al. 2011)
Francisella novicida	Wild type	Nil	Mouse; i.n.	Protective immunity	n.d	(Pierson et al. 2011)
Helicobacter pylori	Wild type	CT	Mouse; i.g., i.n. guinea-pig; i.n.	Protective immunity, IgG	n.d	(Keenan et al. 1998, Keenan et al. 2003)
Neisseria lactamica	Wild type	Nil	Mouse; i.n.,i.p., s.c.	Protective immunity to self and N.m, IgG, IgA	n.d	(Sardinas et al. 2006)
Neisseria meningitidis	Detergent treated	Nil	Rabbit; i.v.	Bactericidal antibodies	OMP	(Zollinger et al. 1978)
Neisseria meningitidis	Wild type	Aluminum hydroxide	Mouse; i.p.	Bactericidal antibodies	LOS	(Zollinger et al. 2010)

(continued)

Table 9.1 (continued)

Organism	OMVs	Adjuvant	Immunization and Route	Immunity	Major antigen	References
Porphyromonas gingivalis	Wild type	Nil	Mouse; i.n.	Protective Immunity; IgG, IgA	n.d	(Nakao et al. 2011)
Salmonella typhimurium	Wild type	Nil	Mouse; i.p.	Protective immunity	n.d	(Alaniz et al. 2007)
Salmonella typhimurium	ΔmsbB; OmpA::CPV VP2 fusion protein	Nil	Mouse; i.p.	Protective immunity, IgG	Recombinantiviral protein VP2	(Lee et al. 2009)
Shigella flexneri	Wild type		Mouse; i.n.,i.d., o, o.c.	Protective immunity, IgG	n.d	(Camacho et al. 2011)
Treponema pallidum	Wild type	Titermax	Mouse; s.c.	Protective immunity	6 OMV antigens	(Blanco et al. 1999)
Vibrio cholerae	Wild type	Nil	Mouse; i.n., i.g., i.p. Rabbit; o	Protective immunity; IgA, IgG	LPS	(Roy et al. 2010, Schild et al. 2008)

i.d.: intradermal, *i.g.*: intragastric, *i.m.*: intramuscular, *i.n.*: intranasal, *i.p.*: intraperitoneal, *i.t.*: intratracheal, *i.v.*: intravenous, *o.c.*: ocular, *o.*: oral, *s.c.*: subcutaneous, *n.d.*: not determined, *N.m*: Neisseria meningitidis, *CT*: Cholera toxin, *LOS*: lipo-oligosaccharide, *LPS*: lipopolysaccharide, *LT*: heat labile enterotoxin, *E TEC*: Enterotoxigenic, *Escherichia coli*

inversely correlated with the presence of serum bactericidal Abs suggesting the importance of antibody-mediated protection against serogroup B infection (Goldschneider et al. 1969). These bacterial Abs were found to be directed against noncapsular antigens (Frasch 1995; Jones and Eldridge 1979). The noncapsular antigens, consisting of the outer membrane proteins and LPS, were therefore considered to be prospective vaccine candidates. The LPS-OMP complex vaccine formulation was found to elicit humoral response and to be protective in mice against challenge with *N. meningitidis* serogroup B. However, when given alone, the purified B polysaccharide induced immunological memory, even though it was incapable of inducing a humoral response (Moreno et al. 1985).

With the partial success of the LPS-OMP complex, attention was drawn to the naturally produced OMVs that bleb off the meningococcal cell surface during growth. These were found to be highly immunogenic (Zollinger et al. 1972) and toxic due to the presence of LOS. The toxicity of OMVs can be reduced by one of several methods that typically involve the use of detergents to remove most of the LOS and phospholipids. Considering the possibility that native OMVs would likely be safe to use via the intranasal (i.n.) route, these were tested for safety and immunogenicity in mice (Saunders et al. 1999). Mice immunized with native OMV vaccine either via the i.n. or intraperitoneal (i.p.) route or by a combination of the two showed that i.n. immunization induced high levels of bactericidal antibody in serum and localized immunoglobulin A (IgA) in the lungs but that i.p. immunization only induced a systemic antibody response (Saunders et al. 1999). Later on, OMVs both native and detergent-treated were used for vaccination in humans and these are discussed in detail in Sect. 9.7.

9.6.2 Neisseria lactamica

N. lactamica is a close relative of *N. meningitidis* and was found to colonize the human nasopharynx, particularly in infants and children. The presence of *N. lactamica* has been implicated in the acquisition of natural immunity to the meningococcus (Gold et al. 1978). Immunological and epidemiological evidence suggests that carriage of *N. lactamica* contributes to the age-related development of natural immunity against the meningococcal disease possibly due to the development of Abs against the many surface structures in common with *N. meningitidis* (Gold et al. 1978; Gorringe 2005).

Like *N. meningitidis* and other Gram-negative organisms, *N. lactamica* cells naturally shed OMVs during growth. The potential of *N. lactamica* OMVs as a vaccine candidate has been examined in a mouse model of meningococcal disease (Gorringe et al. 2005; Oliver et al. 2002). NIH mice were immunized with *N. lactamica* OMVs along with aluminium hydroxide or Freund's complete adjuvant on days 1, 21, and 28. The antibody response was measured at day 35; the Abs raised against OMVs prepared from *N. lactamica* showed a high degree of cross-reactivity against *N. lactamica* when examined by whole-cell enzyme-linked immunosorbent assay (ELISA). For challenge experiments, mice were infected by intraperitoneal injection of *N. meningitidis*. Immunization with *N. lactamica* OMVs was found to protect mice against challenge with all of the meningococcal strains tested (Oliver et al. 2002).

9.6.3 Vibrio cholerae

Vibrio cholerae, the causative agent of the diarrheal disease cholera is a Gram-negative, motile, comma-shaped bacterium. So thus far, although more than 200 serogroups of *V. cholerae* have been identified, only the serogroups O1 and O139 are responsible for the disease (Chaudhuri and Chatterjee 2009; Faruque et al. 1998). Following ingestion through the oral route, *V. cholerae* colonizes the small intestine and secretes the potent enterotoxin, called the cholera toxin, which is primarily responsible for the diarrheal symptoms. The expression of virulence factors is controlled by a complex regulatory cascade composed of ToxT and the inner membrane proteins ToxR/S and TcpP/H (Chaudhuri and Chatterjee 2009). Many of the surface factors associated with the outer membrane, such as LPS, porins, toxin-coregulated pili, and flagella have been demonstrated to be important for colonization of *V. cholerae* in an infant mouse or rabbit ileal loop model of infection (Kaper et al. 1995).

The acute secretory diarrhea caused by *V. cholerae* causes fluid loss and massive dehydration and can lead to death within hours if left untreated (Bennish 1994). The treatment involves simple rehydration therapy, sometimes in combination with antimicrobial agents (Kaper et al. 1995; Sack et al. 2004). The WHO recommends the use of vaccines for cholera in the endemic regions. Currently available safe and effective vaccines are Dukoral and Shancol, both of which are killed whole-cell *V. cholerae* vaccines, one with a B subunit of cholera toxin and the other is without a B subunit (Czerkinsky and Holmgren 2009). Dukoral was developed in Sweden in 1991 and is prequalified and licensed for use in more than 60 countries. It is a monovalent vaccine consisting of formalin and heat-killed whole *V. cholerae* O1 (classical and El Tor, Inaba and Ogawa) cells along with a purified recombinant B subunit of cholera toxin (WC-rBS). Dukoral provides short-term protection (85–90 %) against *V. cholerae* O1 infection. The other vaccine, Shancol, is a bivalent vaccine and consists of killed whole cells from *V. cholerae* O1 and O139. After successful phase II clinical trials in India and Viet Nam, this vaccine was licensed in 2009 as mORCVAX in Viet Nam and as Shanchol in India. It provides longer-term protection compared to Dukoral. However, the WHO recommends the immunization to be used along with control measures as the effects produced by these vaccines are only short term; improving the water supply and having proper sanitation in place will provide a long-term effect. Earlier use of live cholera toxin attenuated vaccines have been stopped due to the potential risk of reversion to a virulent phenotype; efforts are now directed towards developing subunit or acellular vaccines. There is a need for a cholera vaccine that can give long-term protection and is appropriate for field use.

Recently, on the basis of positive examples of immunization with OMVs from a number of Gram-negative bacteria, especially the success of immunization with *N. meningitidis* OMV, and on the report that *V. cholerae* also shreds OMVs during its normal growth (Chatterjee and Das 1966, Chatterjee and Das 1967), the potential of OMV as a vaccine candidate for cholera has been examined (Schild et al. 2009, Schild et al. 2008). In a recent study, female mice were immunized with *V. cholerae* OMVs via intraperitoneal, intragastric, and intranasal routes and the antibody titer was measured in each case. OMVs released from pathogenic *V. cholerae* were purified and concentrated by ultracentrifugation and suspended in phosphate-buffered saline. A high-titer immune response against a variety of antigens present in the OMVs was observed in an infant mouse model independent of the route of immunization (Schild et al. 2008). A protective induced immunity of at least three months was observed in the offspring of immunized female mice. In a later study, it was shown that *V. cholerae* OMVs can be used for expression and delivery of heterologous proteins into a host. The periplasmic alkaline phosphatase (PhoA) of *E. coli* was successfully expressed in *V. cholerae* and incorporated into OMVs. Intranasal immunization with *E. coli* PhoA containing OMVs was found to induce a specific immune response against this heterologous antigen in mice indicating the potential use of *V. cholerae* OMVs as antigen delivery vehicles in vaccine design (Schild et al. 2009). The immune response of OMVs isolated from *V. cholerae* El Tor N16961 was further examined using a removable intestinal tie-adult rabbit diarrhea (RITARD) model (Roy et al. 2010), which is an accepted model closely resembling human cholera. Oral immunization with OMVs provided significant protection against the development of diarrhea in rabbits. Moreover, OMVs protect against the homologous as well as several heterologous strains of *V. cholerae* O1 but were unable to protect significantly against *V. cholerae* O139.

9.6.4 *Acinetobacter baumannii*

Acinetobacter baumannii is a Gram-negative pathogen that produces different types of infections such as pneumonia, meningitis, and bloodstream infections in critically ill patients in intensive care settings or in immune-compromised individuals (Garnacho et al. 2003). Although the organism does not carry specific toxins and is considered to be a low virulent pathogen, the most notable complications in *A. baumannii* infection arise due to the rapid emergence and spread of multidrug-resistant strains. Due to the difficulty in treating infections caused by multidrug-resistant strains, novel approaches are required for prevention of infections caused by this pathogen. Currently no vaccine is available for prevention of *A. baumannii*-induced infection. The use of formalin-inactivated whole cells as a vaccine elicited a robust antibody response and protected mice from infection (McConnell and Pachon 2010). However, the use of inactivated whole cell as vaccine antigen may raise potential safety concerns over the possibility of incomplete activation.

Recent studies have shown that like other organisms, *A. baumannii* secretes OMVs during growth in vitro as well as in vivo (Kwon et al. 2009; Jin et al. 2011). OMVs are a nonviable source of surface antigens, therefore the vaccine potential of OMVs were examined in an animal model and it was expected that the above concern of reversibility could be avoided.

OMVs were isolated from *A. baumannii* strain ATCC 19606, an antibiotic-susceptible reference strain, and examined for its vaccine potential in a mouse model of disseminated sepsis (McConnell et al. 2011). Bacterial cells grown in rich medium were pelleted by centrifugation and the supernatant was filtered through a 0.22-μm filter and ultracentrifuged to collect OMVs. The OMVs were checked for the presence of viable bacteria. OMVs free from live bacteria were suspended in PBS and combined with the aluminum phosphate adjuvant and C57BL/6 mice were immunized by intramuscular injection on days 1 and 14. Immunization resulted in a robust antibody response against multiple bacterial antigens as evidenced by the presence of antigen-specific IgG and IgM. To identify the highly immunogenic OMV proteins, the SDS-PAGE containing OMV proteins were blotted with 21-day serum and the strongly reactive bands were excised and examined by MALDI-TOF/TOF–MS. Six proteins were identified, most of which were found to be located in the bacterial outer membrane. It is important to note that vaccination with OMVs protected mice from challenge with the ATCC 19606 strain and provided protection against two clinical isolates in addition (McConnell et al. 2011). Thus, vaccination with OMVs might be a viable strategy for preventing *A. baumannii* infection.

9.6.5 *Bordetella pertussis*

Bordetella pertussis, fastidious Gram-negative bacteria, is the causative agent of Pertussis or whooping cough, an acute respiratory tract infection. This disease, although preventable by vaccination, remains one of the top ten causes of death worldwide not only among children younger than one year old (Tanaka et al. 2003) but also in adolescents and adults (Ntezayabo et al. 2003). Currently, respiratory disease caused by *B. pertussis* is prevented by administration of an acellular pertussis vaccine (called Pa) composed of up to five antigens: filamentous hemagglutinin (FHA), pertactin (PRN), pertussis toxin (PT), and two fimbrial proteins (Fim) (Locht 2008). Although this vaccine has been proven to be successful at first (70–90 %), subsequently observed vaccine-induced antigenic shifts and adaptations became of concern.

Like other Gram-negative organisms, *Bordetella* also releases OMV which carries a wide spectrum of endogenous antigens and has been used in acellular vaccine development (Roberts et al. 2008). OMVs from the Tohama vaccine strain (CIP 8132), which revealed the presence of the main well-known pertussis immunogens, such as PRN, adenylate cyclase hemolysis, PT, and LOS, were used to examine the protective capacity of OMVs in a murine *B. pertussis* intranasal challenge model. OMVs were isolated from bacterial cells by differential centrifugation, suspended in deoxycholate in Tris EDTA (TE) buffer, and purified by sucrose gradient. Immunization was performed in BALB/c mice via two routes: (1) i.p., with OMVs emulsified with aluminum hydroxide (alum) as an adjuvant and (2) i.n., with OMVs without adjuvant, in a two-dose schedule over two weeks. Following immunization mice were challenged i.n. with sublethal doses of *B. pertussis*. Significant differences in recovery of bacterial colonies between immunized animals and the PBS-treated group were observed ($p < 0.001$) with adequate elimination observed in immunized mice. Thus OMVs of *B. pertussis* could be candidates to be used to protect against pertussis.

In a later study, outer membrane vesicles were engineered to decrease their endotoxicity so that a safer and effective pertussis acelullar vaccine could be designed (Asensio et al. 2011). In order to generate less toxic *B. pertussis* OMVs, a recombinant *B. pertussis* strain containing the *pagL* gene of *B. bronchoseptica* was constructed. The *pagL* gene encodes PagL protein, which is a lipid A modifying enzyme located in the outer membrane and having lipid A 3-O-deacylase activity. This enzyme hydrolyzes the ester bond at the three-position of lipid A, creating tetra-acylated LOS instead of more toxic penta-acylated LOS. This modification modulates the recognition of lipid A by the TLR/MD-2 receptor complex and consequently endotoxic activity (Geurtsen et al. 2007). The protective capacity of PagL expressed OMVs prepared from *B. pertussis* Tohama strains were examined in a *B. pertussis* intranasal challenge model in BALB/c mice. Mice were immunized with OMVs intranasally without adjuvant in a two-dose schedule over 2 weeks. Significant differences in terms of bacterial recoveries in lungs were observed among immunized animals and the PBS- treated group (p < 0.001). Adequate bacterial elimination rates were observed for both wild- type and PagL expressing OMVs suggesting further improvement in safety/efficacy of OMV vaccines through genetic engineering.

9.6.6 *Burkholderia pseudomallei*

Burkholderia pseudomallei, a Gram-negative organism, is an environmental saprophyte endemic in South East Asia and northern Australia (White 2003). It is now recognized as an emerging infectious disease in India (John et al. 1996). When infected by *B. pseudomallei,* usually via a transpercutaneous, respiratory, or oral route, an illness known as melioidosis occurs (Dharakul and Songsivilai 1999). The clinical symptoms of melioidosis can vary, but the most common presenting symptom is severe pulmonary distress, which can eventually progress to frank septicemia and death, if left untreated. *Burkholderia* infections are difficult to treat due to their resistance to multiple antibiotics, biofilm formation, and establishment of intracellular and chronic infection in the host. Although a number of vaccine strategies against *B. pseudomallei* have been explored, no ideal vaccine candidate has yet emerged from preclinical studies (Sarkar-Tyson and Titball 2010).

Inactivated *B. pseudomallei* whole-cell preparations and live-attenuated strains were demonstrated to be highly immunogenic and conferred partial to full protection in murine models (Haque et al. 2006; Sarkar-Tyson and Titball 2010). However, these vaccines are of limited use in humans because of the safety concerns especially for immune-compromised individuals. Alternative approaches include use of purified preparations of LPS, CPS, or protein-based subunit vaccines. All of these provided variable degrees of protection and had not been proved successful. A successful vaccine against *B. pseudomallei*, on the other hand, will likely require the induction of both humoral and cellular-mediated immune (CMI) responses for complete protection and eradication of persistent bacteria and in addition, the vaccine must be safe and efficacious against multiple routes of infection (Healey et al. 2005).

A recent study has investigated the immunogenicity and protective efficacy of *B. pseudomallei*-derived outer membrane vesicles (Nieves et al. 2011). BALB/c mice were immunized with purified preparations of naturally shed *B. pseudomallei* OMVs subcutaneously (s.c) or intranasally at three doses upto 42 days. No additional adjuvant was added to the OMV preparation. One month after the last immunization, the antibody response was determined and a group of mice were challenged with *B. pseudomallei* by aerosol. *B. pseudomallei* OMVs administered subcutaneously (s.c) or intranasally, could induce OMV-specific serum IgG and IgA responses and T-cell memory responses. The OMV- induced IgG responses were found to be generated against multiple protein antigens in the OMV preparation. Following *B. pseudomallei* challenge by aerosol, naïve mice displayed 100 % mortality by day 7, whereas mice immunized s.c with *B. pseudomallei* OMVs were significantly protected against lethal aerosol challenge. The results thus indicate that OMVs provide a vaccine formulation that is able to produce protective humoral and cellular immunity against lethal aerosol challenge in a murine model of melioidosis. Moreover, this vaccine platform constituted a safe and inexpensive immunization strategy against *B. pseudomallei*.

9.6.7 Porphyromonas gingivalis

Porphyromonas gingivalis, a Gram-negative organism, is frequently implicated in chronic and severe adult periodontitis, an oral disease characterized by destruction of periodontal tissues and ultimately exfoliation of the teeth in humans (Lamont and Jenkinson 1998). An association between periodontitis and systemic diseases such as diabetes mellitus, cardiovascular disease, and atherosclerosis has been reported (Desvarieux et al. 2005). Development of a safe vaccine would greatly improve oral and systemic health. Several virulence factors of *P. gingivalis* are known, including fimbriae, gingipains, hemagglutinins, lipopolysaccharides, and outer membrane vesicles (Lamont and Jenkinson 1998).

In the case of *P. gingivalis*, parenteral administration of OMVs from the invasive strain *P. gingivalis* W50 in BALB/c mice was found to be protective against challenge infection (Kesavalu et al. 1992). In a recent study, a *galE* mutant strain of *P. gingivalis* producing few or no outer membrane vesicles was used to demonstrate that OMVs play a significant role in the antigenicity of *P. gingivalis* (Nakao et al. 2011). Mouse antisera was raised against whole cells of the *P. gingivalis* wild- type strain; ELISAs were used to examine the reactivity of these antisera with whole cells of the wild type or the *galE* mutant. The antisera had significantly lower reactivity against the *galE* mutant compared to wild type. OMVs, but not LPS, retained the immunodominant determinant of *P. gingivalis*, as determined by ELISAs (with wild- type LPS or OMVs as antigen) and absorption assays. The intranasal administration of *P. gingivalis* OMVs BALB/c mice elicited dramatically high levels of *P. gingivalis*-specific IgA in nasal washes and saliva, as well as serum IgG and IgA. Synthetic double-stranded RNA polyriboinosinic polyribocytidylic acid [Poly (I: C)], an agonist of Toll-like receptor 3 (TLR3), was used as the mucosal adjuvant. The studies indicated that *P. gingivalis* OMV was an intriguing immunogen for development of a periodontal disease vaccine.

9.6.8 Salmonella typhimurium

Like other Gram-negative pathogens, *Salmonella* can produce OMVs containing outer membrane and periplasmic contents. The studies on the interaction of OMVs with host defenses have raised promises regarding use of OMVs as potential vaccine candidates (Alaniz et al. 2007). First, *Salmonella* OMVs potently stimulate innate proinflammatory responses (secretion of TNF-α, IL-12, and production of NO) from professional antigen-presenting cells (APCs) such as macrophages and DCs. OMVs stimulate DC maturation and activation in TLR4-dependent and independent pathways. Because DCs use uniquely powerful APC functions to efficiently activate adaptive immune responses, DC activation of OMVs is critical in initiating *Salmonella* specific immunity in vivo. In addition, *Salmonella* OMVs express antigens recognized by *Salmonella*-specific CD4$^+$ T cell lines derived from protectively immunized mice (Bergman et al. 2005). These results suggested that Salmonella OMVs might contribute to the development of *Salmonella*-specific adaptive immune responses in vivo. It has been demonstrated that (a) *Salmonella* OMVs contain B- cell and T- cell Ags which influence the development of *Salmonella*-specific immunity to infection in vivo; (b) OMVs, after immunization, prime *Salmonella*-specific immune responses limiting bacterial replication in vivo; (c) OMVs stimulate IFN-γ production by a large proportion of CD4$^+$ T cells from *Salmonella*-immune mice, indicating that OMVs are an abundant source of Ags recognized by Salmonella-specific CD4$^+$ T cells; and (d) naive mice immunized with OMVs develop robust *Salmonella*-specific B-and

T-cell responses (Alaniz et al. 2007). Based on the above results, OMVs of *Salmonella* might be an additional means to induce pathogen-specific immunity, and are therefore attractive vaccine candidates.

For immunization study, OMVs were isolated from the log-phase culture of *Salmonella* pulsed with gentamicin to increase OMV production. After removal of bacterial cells by centrifugation, the supernatant was filtered through a 0.22- μm filter. OMVs were enriched from the supernatant by ammonium sulphate precipitation and differential centrifugation and finally collected by ultracentrifugation. Purified OMVs were reconstituted in sterile nonpyrogenic water for further experiment (Alaniz et al. 2007).

For protection studies, mice were injected with OMVs in phosphate- buffered saline (PBS) at two-week intervals and at a later time OMV-immune mice were infected via the i.p. route with *Salmonella*. After OMV immunization and live bacterial challenge, the bacterial burden in infected tissues was used as a measure of protective immunity. OMV-immune mice had decreased bacterial viability in terms of colony- forming units (CFUs) in all organs measured after infection with live *Salmonella*.

These observations suggest that OMVs possessed protective Ags that were targeted during live infection, and OMVs had intrinsic properties recognized to be important mediators of DC, B, and T cell activation during *Salmonella* infection in vivo, and the studies demonstrate the potential utility of OMVs as a natural complex acellular vaccine candidate (Alaniz et al. 2007).

In a later study, low endotoxic and multi-immunogenic *Salmonella* OMVs containing a foreign epitope were constructed as a potential vaccine candidate (Lee et al. 2009). Endotoxicity of bacterial LPS was critical for vaccine safety and in as much as LPS is an intrinsic part of OMVs, safe vaccines should have OMVs with low endotoxic activities. The endotoxic activity in *Salmonella* is known to be reduced by mutation in the *msbB* gene (Low et al. 1999). OMVs are naturally produced proteoliposomes and have the potential as delivery vehicles for heterologous antigens (Kuehn and Kesty 2005), therefore multi-immunogenic OMVs could serve as a potential broad vaccine candidate for other bacteria and viruses. Foreign proteins, if expressed and secreted into the periplasm or co-localized in the outer membrane via translational fusion with outer membrane proteins, are spontaneously loaded within the OMVs (Kuehn and Kesty 2005; Wang et al. 1999).

To develop low endotoxic and multi-immunogenic OMVs the following strategies were adopted. First, a deletion *msbB* mutant of *S. typhimurium* was constructed that produced OMVs with penta-acylated LPS. Compared to the hexa-acylated wild- type LPS, this structurally modified penta-acylated LPS was known to be much less endotoxic to humans (Low et al. 1999). The low endotoxicity of the *Salmonella msbB* mutant was confirmed by Limulus amebocyte lysate (LAL) assay and mouse lethality test. Second, a recombinant expression vector was constructed to express the fusion protein OmpA::CPV VP2 where a canine parvovirus (CPV) VP2 epitope was fused to the bacterial OmpA protein.

The expression vector containing the fusion protein was transformed to the *Salmonella msbB* deletion mutant.

VP2 is the most abundant capsid protein of canine parvovirus , causative agent of potentially fatal diseases, such as severe gastroenteritis in juvenile dogs and myocarditis in neonatal puppies (Patial et al. 2007). The VP2 protein has been a target antigen to produce neutralizing Abs against CPV. OmpA is one of the most abundant outer membrane proteins in *Salmonella* and confers stability to the outer membrane (Vogel and Jahnig 1986). OmpA was therefore chosen as a fusion target to incorporate the CPV VP2 epitope into the OMV because for incorporation of the said epitope, the fusion protein (OmpA::CPV VP2 epitope) has to be expressed and destined to the outer membrane of the bacteria. In immunogenicity tests, sera obtained from the mice immunized with either the *Salmonella msbB* mutant or its OMVs containing the OmpA::CPV VP2 epitope showed bactericidal activities against wild-type *S. typhimurium* and contained specific Abs to the CPV VP2 epitope. These results suggested the presence of Abs as well as a protective response upon immunization with multi-immunogenic OMVs in BALB/c mice against *Salmonella* infection.

9.6.9 Shigella flexneri

Shigella flexneri, the causative agent of Shigellosis, is responsible for a large number of deaths worldwide due to diarrhea especially among children under the age of five (Kotloff et al. 1999). Shigellosis often causes serious damage of the intestinal epithelium limiting the correct nutrient absorption subsequently leading to a life long condition. Prevention of Shigellosis relies on basic sanitary measures. Furthermore, the increasing problem of antibiotic resistance raises the urgent need of protective vaccines. In fact, the World Health Organization has made the development of a safe and effective vaccine against Shigella a high priority for many years (Kotloff et al. 1999).

Vaccination efforts are mainly focused on live oral vaccines, and although several vaccine candidates are on clinical trial, no safe and efficacious Shigella vaccine is yet available (Kweon 2008). Currently, acellular vaccines are being tried as these are safer in comparison with live-attenuated whole- cell vaccines (Kaminski and Oaks 2009, Levine et al. 2007). However, these vaccines require appropriate adjuvants and researchers face the challenge of designing formulations able to enhance the immunogenicity of associated antigens, through the right activation of the immune system, and susceptible to be administered by mucosal routes. Considering the fact that OMVs naturally produced by *Shigella flexneri* are the storehouse of relevant antigens to be included in the acellular vaccine, recent studies focus on examining the protection conferred by *Shigella* OMV in animal models (Camacho et al. 2011; Mallett et al. 1995). OMVs were isolated and purified from *S. flexneri* 2a, which is the most common cause of Shigellosis and is responsible for 25–50 % of all cases in the developing world (Levine et al. 2007).

OMVs were isolated from the stationary phase culture of *S. flexneri* in a rich medium. After initial centrifugation to remove the cell pellet, the supernatant was filtered through a 0.45- micron polyvinylidene fluoride (PVDF) filter and purified by ultradiafiltration via a 300 kDa tangential filtration concentration unit. Protective efficacy was tested in BALB/c mice immunized with one single dose of OMVs either free or encapsulated in poly methylvinylether-co-maleic anhydride (PVM/MA) nanoparticles. The average size of a nanoparticle- conjugated OMV was approximately 197 nm. Mice were immunized by an intradermal, nasal, ocular, or oral route. Challenge infection was performed on day 35 intranasally with a lethal dose of *S. flexneri* 2a clinical isolate. The number of dead mice after challenge was recorded daily upto 30 days.

Immunization via the nasal or ocular route with free OMVs provided complete protection. The differences in protection of free OMVs and NP-encapsulated OMV immunization groups were nonsignificant when immunized by the nasal, ocular, or oral route. On the other hand, intradermal delivery of free OMVs was not protective, but NP-encapsulated OMVs conferred full protection. Specific IgG2a and IgG1 antibody responses against OMV antigens were determined by indirect-ELISA at postimmunization. Results showed that the OMV immunization by either route elicited significant levels of serum IgG1 and IgG2a with respect to control mice (Camacho et al. 2011).

9.6.10 Treponema pallidum

Treponema pallidum, the causative agent of syphilis, is a noncultivatable spirochete that causes latent infection in both humans and experimental animals. The outer membrane of this organism contains a hundred fold fewer membrane-spanning proteins or porin like proteins compared with outer membranes from typical Gram-negative bacteria (Walker et al. 1989). These membrane- spanning *T. pallidum* rare outer membrane proteins (termed TROMPs) were identified by freeze-fracture analysis (Blanco et al. 1990), and shown to be the only recognized surface- exposed proteins of this organism. The aggregation of the TROMPs was observed following the incubation of *T. pallidum* with serum from syphilitic rabbits immune to challenge reinfection (Blanco et al. 1990).

Evidence suggests that Ab plays a key role in the protective immunity that develops during syphilitic infection in humans and animals (Azadegan et al. 1983; Bishop and Miller 1983). Furthermore, immunization with killed whole organisms, fractionated organisms, or recombinant proteins have, failed to elicit serum treponemicidal activity comparable to that of immune serum.

Considering the fact that there is a surface target of bactericidal antibody, OMVs were tested for immunization studies. BALB/c mice were infected subcutaneously in two sites with OMV derived from *T. pallidum* at 2, 5, and 8 months. Serum was collected by cardiac puncture. Serum was also collected from syphilitic rabbits immune to challenge (immune rabbit serum, IRS). OMVs isolated from *T. pallidum* induced a serum bactericidal activity 30 times greater than that found in IRS (Blanco et al. 1999). Freeze-fracture analysis of *T. pallidum* incubated in IRS or anti-OMV serum showed that *T. pallidum* rare membrane-spanning outer membrane proteins was aggregated, suggesting TROMPs were the targets for serum bactericidal activity.

In a later study, a monoclonal antibody with potent bactericidal activity M131 was isolated following immunization of rabbits with OMVs. This monoclonal antibody bound to a phosphorylcholine epitope on the *T. pallidum* surface and conveyed partial protection in experimental rabbit syphilis following passive immunization.

9.7 OMVs as Vaccine

Infections by Gram-negative organisms still continue to be a major threat worldwide. The treatment strategy, which is mainly the use of antibiotics, is becoming problematic because of increased and rapid emergence of antibiotic-resistant bacteria. Hence vaccination is considered as the most straight forward way to combat bacterial infections. Vaccines are the agents that prime the human body and can induce immunity to a specific infectious disease. Vaccines require two basic elements: (1) an antigen that will trigger immune response and create immunological memory to protect against further attack by the invading organism and (2) an adjuvant to ensure that the response is robust. OMVs are nonviable vesicles naturally released from bacteria and containing a cocktail of immunogenic antigens including lipopolysaccharides and outer membrane proteins making them attractive as a vaccine candidate. These OMVs have been shown to stimulate an adaptive memory immune response and have self-adjuvant properties that can be utilized in the development of effective acellular vaccines. These OMVs can be easily prepared from growing bacterial cultures and can be modified for inclusion of antigens or exclusion of toxic components such as lipo-oligosaccharides. OMV vesicles as vaccines have been extensively studied against serogroup B *Neisseria meningitidis* (Kelly and Rappuoli 2005; Perrett and Pollard 2005; Sadarangani and Pollard 2010; Su and Snape 2011). The concept of OMV- based vaccines has been tried in many other organisms including *Vibrio cholerae, Salmonella typhimurium,* and *Borrelia burgdorferi,* among others, suggesting that OMVs are capable of eliciting stable immune responses and conferring protection against respective bacterial infection in various disease models as described above in Sect. 9.6. The number of studies concentrating on OMVs as vaccine tools is increasing steadily which supports the likelihood of application of OMV vaccines in many other Gram-negative pathogens. A closer look especially at *N. meningitidis* OMV vaccines will provide insight to the prospects and challenges of OMV based vaccine candidates.

9.7.1 OMVs as Vaccines for Meningococcal Diseases

Invasive meningococcal disease is endemic worldwide and caused by *Neisseria meningitidis*. Among 13 different serogroups of this organism, only six (A, B, C, W135, X, and Y) are responsible for the disease (Stephens et al. 2007). Group A is responsible for large epidemics in Africa; groups W-135 and X have emerged in Africa, and group Y in the United States. Groups B and C are predominant in industrialized countries (Stephens et al. 2007; van Deuren et al. 2000). The capsular group B strains cause prolonged epidemics, such as those in Cuba, Norway, and New Zealand (Baker et al. 2001). The highest incidence of disease caused by Group B strains is among infants less than one year of age (Rosenstein et al. 2001; Shepard et al. 2003; Trotter et al. 2007). In the 1960s, bivalent capsular polysaccharide vaccines were developed successfully against the serogroups A and C. Subsequently, CPS vaccines were introduced for serogroups Y and W-135. The tetravalent serogroup vaccine A/C/Y/W135 has been licensed in the United States since 1981 and is in routine clinical use for persons older than 11 years of age (Bilukha and Rosenstein 2005). However, the polysaccharide vaccines had limited utility in childhood vaccination programs due to (a) limited efficacy and (b) short duration of protection. The short span of protection was due to the fact that the polysaccharide antigens induced a T-cell independent immune response thereby causing poor immunological memory. To improve immunological memory, polysaccharide antigens were conjugated to a protein carrier. An Immunization program with this new generation of glycoconjugate vaccines was successful against diseases caused by group C meningococci and was introduced in the United Kingdom and several European countries (EU-IBIS 2002). Later on, quadrivalent serogroup PS vaccine A/C/Y/W135 conjugated to diphtheria toxoid was licensed in the United States (Mitka 2005). A *Hemophilus influenza*e type b-meningococcal Group C and Y is at late-stage clinical development and is suitable for infants (Nolan et al. 2007). None of these PS-based vaccines was effective against diseases caused by *Neisseria meningitidis* serogroup B. Control of meningococcal disease could not be achieved as a broadly effective vaccine was not available. The group B disease remains a major public health concern.

9.7.2 Markers of Protection Against Meningococcal Disease

For a vaccine efficacy trial, surrogate protection markers are needed that can be measured in participants of the trials. Several results suggest that the complement-mediated serum bactericidal antibody (SBA) can be inversely correlated with the incidence of the disease (Goldschneider et al. 1969). SBA was a measure of the ability of postvaccination sera to kill a strain of *N. meningitidis* in the presence of complement. This classic study led to the accepted use of SBA assay as a surrogate marker of protection.

For meningococcal polysaccharide vaccines, commercially available baby rabbit complement was used in a standard bactericidal assay (rSBA;) (Borrow et al. 2005). The criteria of a greater than four fold rise in rSBA titer at two weeks after immunization was used as a correlate of protection in adult recipients of polysaccharide vaccines, whether group A, C, Y, or W135 (Borrow et al. 2005). For serogroup B disease, SBA titers of one-fourth or greater, using human complement, were correlated with efficacy in trials of OMV vaccines and accepted as the protective threshold in assessing experimental vaccines (Holst et al. 2003; Sierra et al. 1991).

9.7.3 Challenges for Vaccine Development

In contrast to other serogroup polysaccharides, group-B polysaccharides were found to be poorly immunogenic even when conjugated with tetanus toxoid (Wyle et al. 1972). The group B capsular structure is composed of $\alpha(2 \rightarrow 8)$ N-acetyl neuraminic acid (polysialic acid) which is structurally identical to fetal brain cell adhesion molecules and is expressed in most of the host tissues (Finne et al. 1983). Attempts were made to overcome this problem by substitution of N-propionyl groups of polysaccharide with N-acetyl groups of the sialic acid residues and then conjugated to tetanus toxoid, but such vaccines although they elicited serum bactericidal response in mice; (Jennings et al. 1987), failed to induce functional antibody response in humans (Bruge et al. 2004; Jennings et al. 1986).

Considerable attention was drawn to the production of noncapsular vaccines and in this venture purified outer membrane proteins were tried. Although promising in animal studies, it failed to induce protective immunity in humans which was attributed to the loss of protein conformation (Frasch and Mocca 1982; Frasch and Peppler 1982; Frasch and Robbins 1978). The principal challenge remaining is to identify safe and antigenically conserved surface exposed noncapsular antigens that can elicit broad SBA responses in humans. Attention was focused on OMVs and several promising approaches involving noncapsular group B OMV vaccines are discussed below.

As do other Gram-negative bacteria, *N. meningitidis* also released so-called blebs or OMVs when grown in liquid culture (Devoe and Gilchrist 1973). Three different formulations have been used by investigators all over the world. Several of these have undergone efficacy trial. These strategies can be broadly classified into: (a) detergent-treated OMVs, (b) Native OMVs, and (c) Recombinant OMVs.

9.7.4 Detergent-Treated OMVs as Vaccine

OMVs were prepared from liquid culture and some of the LPSs were selectively depleted by an appropriate detergent treatment. The detergent was subsequently removed and the proteins were solubilized. During the late 1970s and early 1980s,

Table 9.2 Summary of efficacy trials of OMV-based serogroup B meningococcal vaccines

Location of trial (Reference)	Period of study	Vaccine	Strain	Study design (No. of doses)	Sample size	Age Group	Vaccine efficacy or effectiveness
Cape town, South Africa (Frasch et al. 1983)	1981	OMPs + PS	M986	Randomized double blind (1)	4,400	4 months–5 years	Too few cases
Cuba (Sierra et al. 1991)	1987–1989	VA-MENGOC-BC™	Cu385 (B:4:P1.19,15)	Randomized double blind (2)	106,000	10–14 years	83 % (95 % CI 42–95)
São Paolo, Brazil (de Moraes et al. 1992)	1990–1991	VA-MENGOC-BC™	Cu385 (B:4:P1.19,15)	Case–control	112 cases, 409 controls	<24 months 24–47 months >= 4 years	−37 % (<−100–73) 47 % (−72–84) 74 % (16–92)
Rio de Janeiro, Brazil (Noronha et al. 1995)	1990	VA-MENGOC-BC™	Cu385 (B:4:P1.19,15)	Case–control	275 cases, 279 controls	6–23 months 24–47 months 4–9 years	23 % (−119–73) 42 % (−47–77) 70 % (38–85)
Iquique, Chile (Boslego et al. 1995)	1987–1989	WRAIR	8257, (B:15:P1.3)	Randomized double-blind, placebo controlled	40811	1–4 years 5–21 years	−39 % (<−100–77) 70 % (3–93)
Norway (Bjune et al. 1991b, Bjune et al. 1991a)	1988–1991	MenBVac	44/76 (B:15:P1.7,16)	Randomized double-blind, placebo controlled, cluster randomized efficacy study (2)	171800	13–21 years	57 % (27–87)
New Zealand (Oster et al. 2005, Oster et al. 2007)	2004–2008	MeNZB	NZ98/254 (B:4:P1.7-2.4)	Vaccination campaign (3)	1,032,239	6 months–19 years	73 %

soluble OMV vaccines showed satisfactory safety profiles and immunogenicity, the SBA titer being higher in adults compared to children less than six years age (WHO. 1998). Non-covalent complexing of capsular PS and/or adsorption to aluminum hydroxide (serving as adjuvant) enhanced OMV vaccine immunogenicity. Since then, a number of serogroup B OMV vaccines have undergone efficacy trials with variable results (Table 9.2).

The first efficacy trial of *N. meningitidis* serogroup B OMV vaccine was done at Capetown, South Africa in 1981. In this trial OMVs were extracted from an unencapsulated strain M986 with no adjuvant. Although the vaccine was immunogenic, the study participants were too few to draw any conclusions (Frasch et al. 1983).

The second efficacy trial of OMV based meningococcal vaccine prepared at Finlay Institute of Havana, Cuba (registered trademark VA-MENGOC-BC^{TM}, later commercially marketed as VA-MENGOC-BC$^{®}$) was conducted in Cuba in 1987–1989 (Sierra et al. 1991). OMVs were extracted from a clinical disease isolate (Cu385, B: 4:P1.19, 15) from a local epidemic and by use of a suitable detergent-depleted LOS. The preparation was subsequently enriched with high molecular weight protein complexes and conjugated noncovalently to serogroup C CPS (1:1) that was finally adsorbed to aluminum hydroxide which serves as adjuvant. The trial was double-blind, placebo-controlled, and cluster randomized. About 106,000 students 10–14 years old were enrolled for the study and followed up for 16 months. A good efficacy of 83 % (95 % confidence interval: 42–95) against serogroup B disease was detected (Sierra et al. 1991). There was a decline in incidence of invasive meningococcal disease in Cuba after a mass vaccination campaign with OMV vaccine. The number was 14.3 per 100,000 people in 1983 which declined to 0.8 per 100,000 people in 1993–1994 (Rodriguez et al. 1999).

Two large separate case-control studies with the Cuban vaccine were subsequently conducted in a serogroup B epidemic in Brazil at São Paolo (de Moraes et al. 1992) and Rio de Janeiro (Noronha et al. 1995).The efficacy was 70–74 % in children 4–9 years of age, but was much lower for younger children and absent in infants. The protection was short-lived; there was better protection in the first half-year observation period compared to the subsequent half-year observation period (de Moraes et al. 1992; Noronha et al. 1995).

The next efficacy trial was carried out in Iquique, Chile, from 1987 to 1989 (Boslego et al. 1995) with the vaccine developed at the Walter Reed Army Institute of Research (WRAIR), Washington, DC. The vaccine was prepared from a clinical disease isolate (8257, B:15:P1.3) representative of the Iquique outbreak. The major difference from earlier Cuban vaccine preparation was the use of a zwitterionic detergent, Embigen BB, rather than deoxycholate. Here, the outer membrane proteins were not in the form of vesicles but consisted of multimeric proteins free of LOS. The vaccine contained no class five or other low molecular weight OMPs. The OMPs were noncovalently conjugated with serogroup C capsular PS, and finally absorbed to aluminum hydroxide as adjuvant. The efficacy trial was double-blind, placebo-controlled, and randomized and approximately 40,000 volunteers of ages 1–21 years were enrolled (Boslego et al. 1995). After

20 months of followup a good efficacy of 70 % (95 % CI 3–93) was observed for older children (5–21 years); whereas no protection was detected in younger children (1–4 years; Boslego et al. 1995).

In Norway, a double-blind, placebo-controlled, cluster-randomized efficacy study was conducted in 1988–1991 on around 172,000 secondary school students aged 14–16 years (Bjune et al. 1991a, Bjune et al. 1991b). The study vaccine (MenBvac^{TM}) was developed at the Norwegian Institute of Public Health (NIPH), Oslo, Norway. It contained LOS-depleted OMVs extracted from a clinical disease isolate (44/76, B:15:P1.7,16) representative of the local epidemic and absorbed aluminum hydroxide (Fredriksen et al. 1991). Unlike the Cuban and WRAIR vaccines, it did not contain any meningococcal capsular PS. The efficacy of 57 % was reached after two doses and was considered insufficient. After a relatively long followup period of 29 months, an efficacy of 57 % (lower confidence limit 28 %) was obtained for the public vaccination campaign (Bjune et al. 1991a, Bjune et al. 1991b). There was evidence of better protection during the first phase of observation than the second phase. An efficacy of 87 % was detected in the first 10 months, however, it was as low as 30 % during 21–29 months (Holst et al. 2003). Further studies revealed boosting of protection on application of additional doses. Protection to another subtype of strain was also recorded suggesting broader protection with this vaccine. The Norwegian vaccine was used to control the serogroup B epidemic in Normandy, France that started in 2003. An Efficacy study started in 2006 and the results are still awaited.

Another randomized, double-blind, placebo-controlled trial was carried out with the Norwegian and Cuban OMV vaccines among Icelandic teenagers in 1992–1993 (Perkins et al. 1998). In this study, the main focus was on the evaluation of SBA response (antibody concentrations measured by enzyme immunoassay or EIA) as a potential correlate for vaccine efficacy. Both vaccines had been previously shown to be efficacious for older children and adults in large separate efficacy trials (Bjune et al. 1991a ; Bjune et al. 1991b; Sierra et al. 1991). In contrast to results from these efficacy trials, the proportion of SBA and EIA responders, defined as individuals with a fourfold rise in bactericidal antibody titer or anti-OMV IgG antibody level compared with prevaccination level, was found to be generally lower among the Cuban recipients than the Norwegian vaccine recipients (Perkins et al. 1998). Six weeks after the second dose, 25 and 54 % of the Cuban vaccine recipients, and 71 and 74 % of the Norwegian vaccine recipients showed a response in SBA and EIA, respectively, against their homologous vaccine strains. Based on these results it was concluded that SBA and EIA activities might not be optimal correlates for serogroup B OMV vaccine efficacy (Perkins et al. 1998).

Another issue related to vaccines based on detergent-treated OMVs from a single meningococcal strain that remained to be settled was the specificity of Abs and hence, the extent of protection they might induce. In the Brazilian and Norwegian efficacy trials, no evidence for serosubtype- or strain-restricted protection for the Cuban and the Norwegian OMV vaccines was detected (de Moraes et al. 1992; Wedege et al. 1999), suggesting that both vaccines can provide some protection against heterologous strains as well. In line with these results, in the Chilean and Icelandic immunogenicity trials, adults, teenagers, and children aged two to four years were able to develop SBA responses against both the homologous (vaccine type) and heterologous strains (Perkins et al. 1998; Tappero et al. 1999) although much lower responses were detected against heterologous strains. This was in contrast to infants younger than one year among whom the SBA responses were mainly directed towards PorA protein and no postvaccination SBA response against the heterologous strains was detected (Tappero et al. 1999).

Another serogroup B meningococcal vaccine was licensed after the Cuban vaccine was used in New Zealand. New Zealand experienced a prolonged and intense epidemic of group B meningococcal disease that started in 1991 and reached a peak in 2001 (Baker et al. 2001). A "tailor-made" OMV vaccine (MeNZB^{TM}; Holst et al. 2005; Oster et al. 2005) had undergone a single-center, randomized, observer-blind clinical trial in New Zealand (Ameratunga et al. 2005) to evaluate the safety, tolerability, and immunogenicity in an age group ranging from 6 months to 50 years. The vaccine was produced in collaboration of NIPH with Chiron (Sienna, Italy) from a local epidemic strain B:4:P1.7b,4 (Martin et al. 1998). This subtype accounted for 86 % of all serogroup B meningococci isolated from disease cases from 1990 to 2003 (Martin and McDowell 2004). The vaccine was prepared by fermentor growth of the strain followed by inactivation and extraction of the OMPs with the detergent deoxycholate. Intact fragments of OMVs containing OMP and lipopolysaccharide were purified by ultracentrifugation and adsorbed onto aluminum hydroxide. MeNZB^{TM} was administered in a three-dose regimen by intramuscular injection. An estimate of vaccine effectiveness had been provided. The vaccine was well tolerated, and induced a fourfold rise in serum bactericidal Abs against the vaccine strain 4–6 weeks after the third vaccination in 96 % of adults, 76 % of children, 75 % of toddlers, and 74 % of infants. In a followup study, 6 to 10-week old infants showed a fourfold increase in bactericidal antibody titer after administration of the fourth dose (Oster et al. 2007). MeNZB^{TM} vaccine was offered to all New Zealand babies, children, and teenagers from 2004 to 2008. Since 2008 the vaccination program has no longer been in New Zealand's immunization schedule. The rationale behind this is the fact that the vaccine protected only against the strain from which it had been designed and not any other type or strain of meningococcal disease. The rates of disease (http://www.moh.govt.nz/moh.nsf/indexmh/immunisation-diseasesand vaccines-meningococcal disease) caused by this particular strain fell to a level where experts advised that offering it routinely was no longer necessary.

One limiting criterion for OMV-based vaccine was the fact that this vaccine generated a largely strain-specific immune response, mainly against the protein PorA. This PorA protein was highly variable across strains. There were about 600 PorA variants of which a few have been associated with the disease isolates (Urwin et al. 2004). Thus the vaccines were successful in epidemics caused by a single PorA expressing strain, but were ineffective against a diverse range of PorA proteins found across strains that caused endemic disease. The degree of PorA divergence would make it impractical to use unmodified OMV vaccine.

Furthermore, the protection seemed relatively short-lived (Holst et al. 2003, Noronha et al. 1995); infants required a fourth dose of MeNZB^{TM} during the New Zealand vaccination programme.

9.7.5 Native OMVs as Vaccine

The outcome of the deoxycholate extracted OMV vaccines in human trials as described above were variable: 50–83 % efficacy was obtained in children. These vaccines have been used in specific epidemics and were relatively efficacious when used in the context of an epidemic involving a single bacteria clone. OMVs were detergent-treated in order to lower the endotoxin activity of OMV vaccines. However, detergent treatment may (a) alter the conformation of OMVs, (b) expose epitopes that are not surface exposed under normal conditions, or (c) extract desirable antigens such as lipoproteins. Prepared vaccines may therefore have reduced capacity to induce bactericidal Abs. Another limitation for detergent-treated OMV vaccines is that the serum bactericidal responses in children were found to be largely directed towards PorA protein which is antigenically variable and might therefore be protective against a specific strain (Holst 2007; Holst et al. 2005; Tondella et al. 2000).

An alternative approach has been developed where the vaccine is based on intact membrane vesicles (native OMVs) that have not been exposed to detergents or denaturing solvents. Native OMVs are released into the growth medium during normal growth of bacteria. These native OMVs consist of intact outer membrane antigens in their native conformation and in the membrane environment. These OMVs contain LOS but periplasmic components are not exposed due to the absence of detergent treatment.

Immunogenicity and safety trial of a native OMV prepared from a capsule negative strain of *Neisseria meningitidis* showed that native OMVs are safe intranasal vaccines and were capable of inducing both local and systemic antibody production (Drabick et al. 1999; Katial et al. 2002). The vaccine strain was prepared from *Neisseria meningitidis* 9162 (B:15:P1.3), a clinical isolate from a patient in Chile. The 9162 strain was genetically engineered to delete *synX*. The *synX* gene codes for an enzyme involved in biosynthesis of sialic acid, which is required for sialylation of LPS and capsule formation. The modification resulted in a capsule negative strain containing nonsialylated LPS. Using intranasal native OMV vaccine, successful induction of bactericidal Abs was observed against PorA and L3,7,9 LPS in humans (Drabick et al. 1999). The trial used two different doses of vaccine, all delivered intranasally, the systemic response of which was measured with bactericidal assay and enzyme-linked immunosorbent assay. Persistent bactericidal Abs ($\geq$ fourfold increase) were produced in 75 % of the recipients. In addition to systemic response, the vaccine also produced a local antibody response as detected in the nasal wash fluid of volunteers. In another trial of the same vaccine but with a different group of volunteers and a narrower dosing interval, it was found that the intranasal vaccination with native OMV was safe and could induce both systemic and local immune response (Katial et al. 2002). The vaccine was well tolerated without evidence of inflammation on nasal cytology. Eighteen of 42 volunteers demonstrated a fourfold or greater rise in bactericidal titers, with 81 % showing an increase over baseline.

Although studies had proved that native OMVs were safe intranasal vaccines, the LOS content exceeded the amounts previously shown to be safe for parenteral vaccines. Native OMV vaccines were therefore prepared from strains that had naturally low endotoxic activity (Fransen et al. 2009) or attenuated endotoxic activity by genetic manipulation (Fisseha et al. 2005; Koeberling et al. 2009; Koeberling et al. 2008). The toxicity of LPS of Gram-negative bacteria had been well studied and the lipid A portion of Neisserial LOS was responsible for the endotoxic activity. In wild-type *N. meningitidis*, the lipid A molecule contains six fatty acid chains symmetrically attached via acyl linkages to a $\beta(1' \rightarrow 6)$-linked D-glucosamine disaccharide, phosphorylated at the C1 of the alpha-configured glucosamine and at the C4′ position of the beta-configured glucosamine (Kulshin et al. 1992). In vitro studies have shown that in addition to the extent of phosphorylation, the amount of lipid A acylation as well as the nature of acylation or the length of the fatty acid chains influenced the endotoxic activity. In addition, each glucosamine residue was substituted with a 14:0 (3-OH) amide-linked fatty acid at the 2-position and with a 12:0 (3-OH) ester-linked fatty acid at the 3-position. The hydroxyl groups of the amide-linked fatty acids are acylated with 12:0 fatty acids, and these acyloxyacyl-linked fatty acids were important for the toxicity of lipid A. *N. meningitidis* mutant with deletion in the *lpxl* (*msbB* equivalent of *E. coli*) gene had attenuated endotoxic activity (Weynants et al. 2009). Instead of toxic hexa-acylated lipid A found in wild-type strains, the mutant formed penta-acylated lipid A, which is poorly recognized by human Toll-like receptor 4, the normal LPS receptor (Steeghs et al. 2008).

To enhance immunogenicity, native OMV vaccines were prepared from mutants engineered to overexpress desired antigens such as factor H binding protein (fHbp; Hou et al. 2005; Koeberling et al. 2009; Koeberling et al. 2008). Factor H is an inhibitor of the complement alternative pathway. Binding of factor H to fHBP enables *N. meningitidis* to evade killing by the innate immune system. Mice immunized with a native OMV vaccine prepared from genetically detoxified endotoxin and overexpressed fHbp strain showed high titers of serum anti-fHbp Abs along with broad SBA. The majority of the bactericidal Abs were found to be directed at fHbp by adsorption studies (Koeberling et al. 2009; Koeberling et al. 2008). In addition, strain-specific bactericidal anti-PorA Abs were also found (Koeberling et al. 2009). The safety and immunogenicity of a prototype native OMV vaccine from a mutant *N. meningitidis* strain with genetically attenuated endotoxic activity, overexpressed fHbp, more than one PorA variable region (VR) type, and other mutations were being investigated in a phase 1 clinical trial involving adults (Zollinger et al. 2008).

As anti-LOS Abs were found to show serum bactericidal activity and/or opsonic activity (Cox et al. 2005; Estabrook et al. 2007; Plested et al. 2003), LOS could be a potential meningococcal vaccine target. However, the lacto-n-neotetraose (Gal-GlcNAc-Gal-Glc tetrasaccharide) on meningococcal LOS is shared by antigens on human red blood cells, raising safety concerns. In addition, detergent treatments of OMV vaccines in order to extract LOS decreased LOS immunogenicity (Koeberling et al. 2008; Weynants et al. 2009). To develop an immunogenic LOS-enriched OMV vaccine, Weynants and co-workers prepared an LpxL1 KO mutant with attenuated endotoxic activity (Weynants et al. 2009). To eliminate antigenic cross-reactivity with red cell antigens, the mutant was further engineered to express a truncated LOS lacking the terminal galactose of lacto-n-neotetraose. In addition, PorA was deleted to avoid the hypothetical possibility of PorA immunodominance and suppression of anti-LOS antibody responses. In mice, the mutant OMV vaccine, containing only ∼15 % LOS after mild detergent treatment, elicited high serum anti-LOS antibody titers with broad bactericidal activity. The SBA responses, however, were measured with rabbit complement, which gave much higher titer than did human complement (Santos et al. 2001; Zollinger and Mandrell 1983). One reason could be that rabbit complement factor H (fH) binds poorly to *N. meningitidis* (Granoff et al. 2009). In other studies, mice that were administered native meningococcal OMV vaccines also developed high titers of anti-LOS Abs, as determined by enzyme-linked immunosorbent assay, but the Abs appeared to have low avidity (Koeberling et al. 2008) and had minimal SBA when assayed with human complement (Koeberling et al. 2009; Koeberling et al. 2008; Moe et al. 2002). Thus, the vaccine potential of vaccines that targeted LOS antigens was unknown.

9.7.6 OMV Vaccines in Combination

Efforts to develop a broadly protective *Neisseria meningitidis* group B vaccine had led to many different approaches, some of which have already been discussed. These included detergent-treated OMVs and native OMVs of wild-type and mutant strains. The detergent-treated OMVs had limited coverage because the antibody response was largely mediated towards PorA protein, which is a major constituent of OMVs and is antigenic. Selective removal of LPS decreases toxicity but promotes aggregation resulting in a narrower immune response. Native OMVs retain all LPS and preserve the vesicle structure but result in high toxicity and lower yield. To overcome this limitation, vaccines were engineered based on two or more OMVs, each containing multiple PorA proteins (Cartwright et al. 1999). A hexavalent PorA OMV vaccine (HexaMen; Claassen et al. 1996) was developed at the Netherlands Vaccine Institute (NVI), formerly the National Institute of Public Health and the Environment, RIVM, in the Netherlands. The aim of the vaccine strain construction was to improve the range of protection and at the same time to remove unwanted components. This vaccine was composed of OMVs from two genetically engineered trivalent strains, PL16215 and PL10124, each expressing three different PorA proteins (P1.5,2; P1.7,16; P1.19,15 and P1.7 h,4; P1.5 c,10;

P1.12,13, respectively; Claassen et al. 1996), together covering approximately 80 % of the prevalent strains in the United Kingdom (Cartwright et al. 1999). The nomenclature of PorA serosubtypes were based on two variable regions VR1 and VR2 of PorA. The major structural differences in VR1 and VR2 generate two separate subtype-specific antigenic determinants and form the basis of the sero-subtyping (Frasch and Mocca 1982). In this scheme PorA is assigned by the prefix "P1″ followed by numbers separated by commas that describe the subtype designation. For example, P1.7, 16 has subtype determinants P1.7 and P1.16 in the VR1 and VR2, respectively. Purified vesicles from these strains were formulated into vaccine preparations. The vaccines were devoid of CPS, lacto-N-neotetraose and PorB OMP through gene deletion and expressed only a small amount of class 4 and class 5 OMPs. Lacto-N-neotetraose structure present in Neisserial LPS was also found in human glycolipids; its lack of expression in a vaccine strain was therefore desirable. Class 4 OMPs are encoded by the *rmpM* gene; their possible function is to induce blocking Abs. Class 5 OMPs are encoded by four *opa* genes and show a high degree of inter- and intrastrain variation. The vaccine consisted of 2.57–10 % LPS relative to the protein content, absorbed to aluminum phosphate, and contained sucrose as a stabilizer. The entire engineering was accomplished through transformation with plasmid constructs made in *Escherichia coli* and their homologous recombination into the meningococcal chromosome.

Clinical phase I and II studies in The Netherlands (de Kleijn et al. 2000) and the United Kingdom (Cartwright et al. 1999) had shown the vaccine to be safe and immunogenic in infants, toddlers, and school children. SBA responses have been detected against all of the six PorA subtypes included in the vaccine. However, multiple doses of vaccine were required to induce a significant rise in SBA response and differences were found in the magnitudes of SBA responses to different PorA proteins, suggesting intrinsic differences in the immunogenicity among different PorA subtypes.

To provide even broader protection, a nonavalent OMV vaccine has been developed by addition of a third OMV containing three other PorA subtypes. The vaccine thus contained the nine most common in the industrialized countries, and was developed in collaboration of NVI with Wyeth (New Jersey, United States). The vaccine elicited SBA in mice against most of the targeted PorA variants and in clinical trials. The hexavalent vaccine represented 60–70 % of the meningococcal disease coverage whereas with nonavalent vaccine coverage of 70–89 % could be achieved (Trotter and Ramsay 2007).

One limitation of OMV vaccines is that SBA responses in children are largely directed against PorA protein which is antigenically variable. The OMV vaccines are therefore suitable for epidemics caused by a single clone. To broaden protection, OMV vaccines have been prepared from more than one strain or strains engineered to express more than one PorA protein.

Two other studies were conducted based on the combination of two OMV vaccines or bivalent vaccine. These studies were carried out with the combination of Cuban OMV vaccine MenBVac with MeNZB or New Zealand OMV vaccine. MenBVac was manufactured at NIPH; OMVs were extracted with the detergent

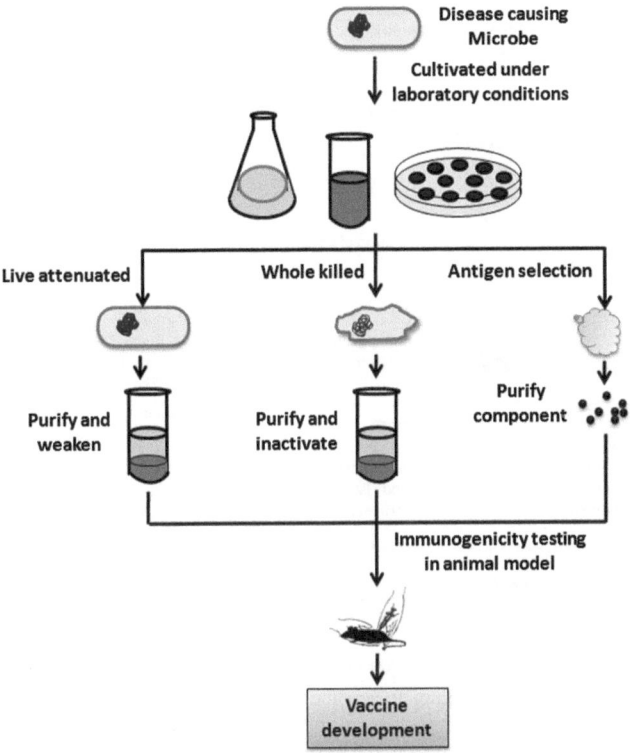

Fig. 9.5 Schematic representation of the procedure for conventional vaccine development

deoxycholate, purified by fractional centrifugation, and adsorbed onto aluminum hydroxide. MeNZB was prepared at NIPH similarly. The combination vaccine was prepared by mixing MenBVac and MenZB immediately before injection. One dose (0.5 ml) of the combination vaccine contained MenBVac (12.5 μg) and MeNZB (12.5 μg) and the same amount of aluminum hydroxide as either of the two separate monovalent vaccines. In the study conducted at Oslo, Norway, an SBA of fourfold or greater was detected in 68–87 % of adults. In the randomized controlled phase II study in Spain and Belgium, this bivalent vaccine yielded fourfold or more SBA in 42–76 % of the participants of age group 12–18 years (Boutriau et al. 2007).

Use of a combination of the two safe and efficacious vaccines MenB and MenZB a new experimental approach is one way to increase protection against serogroup B meningococcal disease when different strains are circulating. These two vaccines will cover the most commonly occurring Serogroup B strains in many European countries. The results from primary immunization showed that the combined vaccine was immunogenic with respect to both vaccine strains and the responses were of the same magnitude as observed for each individual vaccine.

Later, a new experimental approach was described in order to broaden the immunity of the conventional OMV vaccines. The approach consisted of

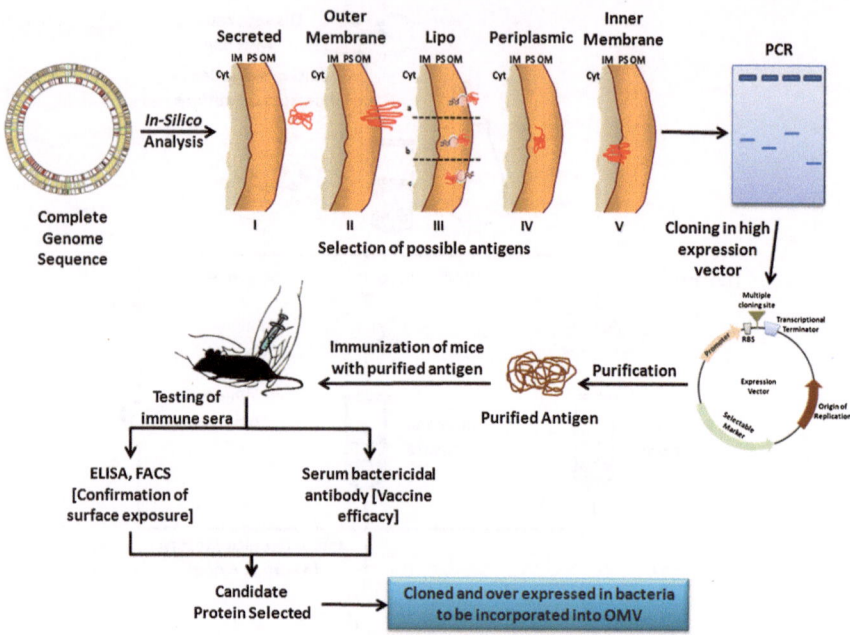

Fig. 9.6 Procedure for recombinant OMV vaccine development in *N. meningitidis* by reverse vaccinology. Starting from the complete genome sequence, in silico analysis identified potential vaccine candidates based on their cellular localization. Selected ORFs were amplified by PCR, cloned in expression vectors, expressed and purified, and were used for immunization of mice. Immune sera were analyzed by FACS to verify whether the antigens are expressed and surface-exposed and also for safety and bactericidal activity. The selected candidate proteins were cloned and expressed in *N. meningitidis* to be incorporated into naturally produced MVs

sequential immunization with three doses of microvesicles (MVs) or OMVs prepared from three antigenically different strains rather than one meningococcal strain (Moe et al. 2002). The rationale behind this approach was that sequential immunization with OMVs prepared from antigenically different meningococcal strains would direct the immune response towards two qualitatively different antibody populations leading to broader protection.

9.7.7 Recombinant OMV Vaccines Using Reverse Vaccinology Approach

Recent alternative approaches involve combination of OMVs with novel vaccine candidates that are mostly noncapsular antigens. As genome sequences became available, a new concept of vaccine design came into play. Bioinformatics tools can predict the function and putative cellular localization of any open reading frame in the genome. These can therefore select the molecules that are likely more

Table 9.3 Potential *Neisseria meningitidis* vaccine antigens identified by "reverse vaccinology"

Protein (molecular mass)	Function	Location	Conserved or not	Vaccine potential	Reference
GNA 1870 (26.9 kDa)	A new lipoprotein	Surface exposed, detected in OMVs	>62 % conservation among 3 variants	High serum bactericidal antibody (SBA) in infant rat model	(Masignani et al. 2003b)
App (160 kDa)	Autotransporter, involved in secretion; binds to epithelial cells	Secreted/surface exposed	Conserved among meningococci	Immunogenic in humans	(Serruto et al. 2003)
NadA (36 kDa)	Adhesin, can bind to human epithelial cells	Bacterial outer membrane	Conserved in hypervirulent lineages	Strong SBA resonse in infant rat model	(Comanducci et al. 2002)
GNA33 (48 kDa)	Murein lytic transglycosylase	Membrane bound	Well conserved among serogroup B	Elicits bactericidal activity in infant rat model	(Jennings et al. 2002)
GNA992 (62 kDa)	Trimeric autotransporter adhesin	Outer membrane protein	Well conserved	Strong SBA in mice	(Pizza et al. 2000)
GNA2132 (51 kDa)	Lipoprotein, heparin binding	Surface exposed	Conserved in diverse strains	SBA in mice and human	(Serruto et al. 2010)
NarE(16 kDa)	ADP-ribosylating enzyme	Periplasmic	Well conserved	N.d	(Masignani et al. 2003a)

GNA: Genome-derived *Neisseria* Antigen, *App*: Adhesion and Penetration Protein, *NadA*: *Neisseria* Adhesion A

effective vaccine candidates, regardless of their abundance or if they are expressed under in vitro or in vivo conditions. This revolutionary approach in vaccine research is termed "reverse vaccinology," as it is different from conventional vaccine design (Fig. 9.5) involving attenuation of the pathogen, inactivation of the micro-organism or creation of subunit preparations.

The conventional approaches have been successful in development of vaccines against a number of infectious diseases (Kelly and Rappuoli 2005). Identification of protective antigen is a pre-requisite for subunit vaccination. The pathogen is cultivated under laboratory conditions and the components that are important for pathogenesis and immunity are identified by biochemical, serological, or genetic methods. The identified protective antigens are then purified and produced in large scale, often by recombinant protein expression. However, the method is time consuming and fails to provide vaccine in many human pathogens. One practical limitation is that it relies on the identification of abundant or immunodominant antigens that may not be the best target for induction of protection.

Reverse vaccinology (Fig 9.6), on the other hand, starts with the genome sequence data, uses in silico analysis to screen genomes for surface-exposed proteins or putative antigens, high-throughput expression analysis by DNA microarray and proteomics analysis, experimental validation of surface location, animal-based immunogenicity testing, and finally conventional safety and efficacy trials in humans. Reverse vaccinology has led to the discovery of potential vaccine candidates that were not possible in earlier studies. The first genome sequence of a serogroup B meningococcal strain MC58 was available in 2000 (Tettelin et al. 2000). A plethora of genes encoding new meningococcal B vaccine candidates have been identified through bioinformatic analysis of the genome sequence and these were then further tested for their vaccine potential in model systems. (Beernink et al. 2006; Parkhill et al. 2000; Pizza et al. 2000; Tettelin et al. 2000). A few of these vaccine candidates are listed in Table 9.3.

Among these, GNA1870 has already been tested for protective response in the OMV vaccine platform. OMVs were prepared from a *N. meningitidis* strain genetically engineered to overexpress GNA1870 variant 1 protein. The modified GNA1870-OMV vaccine elicited broader protection against meningococcal disease than recombinant GNA1870 protein or conventional OMV vaccine in an infant rat model (Hou et al. 2005). The region 100–254 amino acid of GNA1870 has been demonstrated to be important for eliciting bactericidal responses (Giuliani et al. 2005). GNA1870 protein anchored within OMV could provide appropriate conformation of the desired epitopes than in the corresponding portion of recombinant protein vaccine.

Researchers are combining proteomic analysis, functional genomics, and structural vaccinology to identify target antigens as they probe to exploit the "OMV concept" to formulate cross-protective vesicle vaccines (Holst et al. 2009). With the accumulation of data on protective antigens and increased understanding about the mechanism of vesiculation, widespread use of OMVs in vaccine development is expected along with the expansion of global vaccine strategies.

9.7.8 OMVs in Vaccine Delivery

Among vaccines, the subunit vaccines, especially those consisting of protein antigens are the most attractive vaccine design in recent times, as they induce a protective immune response and avoid the safety concerns of the live attenuated vaccines or killed whole-cell vaccines. However, to enhance their immunogenicity, subunit antigens are often combined with appropriate adjuvants (O'Hagan and Valiante 2003), conjugated to protein or a polysaccharide carrier (Koser et al. 2004; Wu et al. 2006), or use formulations in a controlled release system. The latter involves polymers, immunostimulatory complexes, or liposomes, proteosomes, or related vesicles which when combined with the antigen can produce the desired immune response at a single dose (Lowell et al. 1988, Morein et al. 1984; Singh et al. 2007). OMV-derived vaccines contain several naturally produced antigens and apart from this, OMVs find applications as adjuvants in vaccine formulations. Because OMVs are nanoscale proteoliposomes that are released by bacteria, they form attractive candidates for delivery of vaccines too. Recently, OMVs have been engineered to be used as delivery vehicles (see Sect. 9.7.10). These two applications are discussed below.

9.7.9 OMVs as Adjuvant

An adjuvant is an immune-stimulatory agent used in combination with a vaccine to increase the antigen-specific immune response of the vaccine, but is itself free from eliciting a specific antigenic effect. Mucosal vaccines consisting of live attenuated microbes are well known to induce mucosal and systemic immune responses and are highly protective (Dietrich et al. 2003), but at the same time they are accompanied by risk, especially for immune-compromised patients. Therefore the desire for nonreplicating vaccines is rising day by day. Protein vaccines administered by the intranasal route do not tend to induce strong immune responses but these responses have been improved through the use of mucosal adjuvants such as cholera toxin or *E. coli* heat-labile toxin (Pizza et al. 2001). The challenge for developing new and improved adjuvants stems from the need for more potent vaccines and reducing reactogenicity and side effects and also to gain features such as antigen-sparing ability, more rapid protection, stimulation of T-cell immunity, and long-lasting protective immunity.

The outer membrane vesicle (OMV) of *Neisseria meningitidis* was shown to be a potential carrier (conjugated to a hapten) to promote T-cell dependent responses (Sharifat Salmani et al. 2009; Siadat et al. 2007).The mucosal and systemic antibody responses to an influenza virus vaccine are greatly augmented when co-administered with OMV (Haneberg et al. 1998). In addition, the adjuvant properties of OMV-derived particles have been demonstrated for potential cancer vaccines (Estevez et al. 1999). It has been shown that the predominant outer membrane proteins (OMPs; PorA, PorB, and RmpM) from *N. meningitidis* present in the Meningococci B Cuban vaccine had differing capacities to prime the immune system. In addition, OMV isolated from commensal *Neisseria* can be a delivery vehicle for heterologous vaccine antigens. Sardinas et al. (Sardinas et al. 2006) have successfully employed a hepatitis B surface antigen (HBsAg) vaccine as a model antigen to assess the adjuvant properties of *N. lactamica* OMV which leads to increased titer of HBsAb (Sardinas et al. 2006). *N. meningitidis* serogroup B OMV complexed noncovalently with group A meningococcal capsular polysaccharide (GAMP) has been able to induce a high level of bactericidal antibody and opsonophagocytosis activity response in comparison with GAMP alone when tested in New Zealand white rabbits (Siadat et al. 2011). Thus OMVs from either pathogenic or commensal *Neisseria* species can act as effective adjuvants.

9.7.10 OMV as Controlled Release System for Vaccine

OMVs are nanosized (~ 100 nm) proteoliposomes that constitutively bleb out of the outer membrane of Gram-negative bacteria, and they contain the components derived from the bacterial outer membrane and periplasm. It has been observed that some membrane or periplasmic proteins are enriched in the vesicles and some others may be missing. The mechanism of enrichment/exclusion is poorly understood. However, it is clear that some sorting mechanism might be operative coordinating the selective exclusion/enrichment. This phenomenon has been exploited to widen the application of OMV vaccines by using them as delivery vehicles of heterologous antigens (Chen et al. 2010).

ClyA (also referred to as SheA or HlyE) is a bacterial hemolysin that is specifically enriched in OMVs of pathogenic and nonpathogenic *E. coli* and *Salmonella* enteric serovar Typhi and Paratyphi (Kouokam and Wai 2006; Wai et al. 2003). The ClyA protein has an intrinsic ability to translocate across the cytoplasmic membrane without cleavage of the N-terminal signal sequence (del Castillo et al. 2001). Furthermore, it is surface exposed and released from the bacterial cells. Further study revealed that the ClyA cytotoxin is exported from the bacterial cell in OMVs and it adopts a cytolytically active, oligomeric conformation in the vesicles (Wai et al. 2003). Thus the study demonstrates the presence of a vesicle-mediated transport mechanism in bacteria that could be responsible for delivery of pathogenic effector proteins directly into the host cells.

Genetic fusion of ClyA with GFP creating chimera of ClyA–GFP resulted in efficient translocation of heterologous proteins to OMVs (Kim et al. 2008). It was further demonstrated that direct fusion of β-lactamase (Bla), organophosphorus hydrolase (OPH), green fluorescent protein (GFP), and antidigoxin single-chain antibody fragment (scFv.Dig) to the C terminus of ClyA resulted in functional display of each protein on the surface of *E. coli* cells and their derived OMVs. Additionally, Galen et al. (2004) showed that antigens fused to ClyA exhibit immunostimulatory activity when secreted from a live Salmonella vaccine. In a later study, BALB/c mice were immunized with GFP, empty OMVs mixed with ClyA–GFP, or engineered OMVs containing the ClyA–GFP fusion protein. The OMV containing Cly–GFP chimera elicited a strong immune response compared to GFP alone. The anti-GFP humoral response in mice immunized with the engineered OMV formulations was indistinguishable from the response to the purified ClyA–GFP fusion protein alone and equal to purified proteins absorbed to aluminum hydroxide, a standard adjuvant (Chen et al. 2010). With the diverse collection of heterologous proteins that can be functionally localized with OMVs when fused with ClyA, this work signals the possibility of OMVs as a robust and tunable technology platform for a new generation of prophylactic and therapeutic vaccines.

References

Alaniz RC, Deatherage BL, Lara JC, Cookson BT (2007) Membrane vesicles are immunogenic facsimiles of Salmonella typhimurium that potently activate dendritic cells, prime B and T cell responses, and stimulate protective immunity in vivo. J Immunol 179:7692–7701

Ameratunga S, Macmillan A, Stewart J, Scott D, Mulholland K, Crengle S (2005) Evaluating the post-licensure effectiveness of a group B meningococcal vaccine in New Zealand: a multi-faceted strategy. Vaccine 23:2231–2234

Asensio CJ, Gaillard ME, Moreno G, Bottero D, Zurita E et al (2011) Outer membrane vesicles obtained from Bordetella pertussis Tohama expressing the lipid A deacylase PagL as a novel acellular vaccine candidate. Vaccine 29:1649–1656

Attia AS, Ram S, Rice PA, Hansen EJ (2006) Binding of vitronectin by the Moraxella catarrhalis UspA2 protein interferes with late stages of the complement cascade. Infect Immun 74:1597–1611

Azadegan AA, Schell RF, LeFrock JL (1983) Immune serum confers protection against syphilitic infection on hamsters. Infect Immun 42:42–47

Baker MG, Martin DR, Kieft CE, Lennon D (2001) A 10-year serogroup B meningococcal disease epidemic in New Zealand: descriptive epidemiology, 1991–2000. J Paediatr Child Health 37:S13–S19

Bauman SJ, Kuehn MJ (2009) Pseudomonas aeruginosa vesicles associate with and are internalized by human lung epithelial cells. BMC Microbiol 9:26

Beernink PT, Leipus A, Granoff DM (2006) Rapid genetic grouping of factor h-binding protein (genome-derived Neisserial antigen 1870), a promising group B meningococcal vaccine candidate. Clin Vaccine Immunol 13:758–763

Bennish ML (1994) Cholera: pathophysiology, clinical features, and treatment. In: Wachsmuth KI, Blake PA, Olsik O (eds) Vibrio cholerae and cholera: molecular to global perspectives. ASM Press, Washington, DC

Bergman MA, Cummings LA, Barrett SL, Smith KD, Lara JC, Aderem A, Cookson BT (2005) CD4 + T cells and toll-like receptors recognize Salmonella antigens expressed in bacterial surface organelles. Infect Immun 73:1350–1356

Beveridge TJ (1999) Structures of gram-negative cell walls and their derived membrane vesicles. J Bacteriol 181:4725–4733

Bilukha OO, Rosenstein N (2005) Prevention and control of meningococcal disease. Recommendations of the advisory committee on immunization practices (ACIP). MMWR Recomm Rep 54:1–21

Bishop NH, Miller JN (1983) Humoral immune mechanisms in acquired syphilis. In: Schell RF, Musher DM (eds) Pathogenesis and immunology of treponemal infection. Marcel Dekker, New York

Bjerre A, Brusletto B, Mollnes TE, Fritzsonn E, Rosenqvist E et al (2002) Complement activation induced by purified Neisseria meningitidis lipopolysaccharide (LPS), outer membrane vesicles, whole bacteria, and an LPS-free mutant. J Infect Dis 185:220–228

Bjune G, Gronnesby JK, Hoiby EA, Closs O, Nokleby H (1991a) Results of an efficacy trial with an outer membrane vesicle vaccine against systemic serogroup B meningococcal disease in Norway. NIPH Ann 14:125–130; discussion 130–122

Bjune G, Hoiby EA, Gronnesby JK, Arnesen O, Fredriksen JH et al (1991b) Effect of outer membrane vesicle vaccine against group B meningococcal disease in Norway. Lancet 338:1093–1096

Blanco DR, Walker EM, Haake DA, Champion CI, Miller JN, Lovett MA (1990) Complement activation limits the rate of in vitro treponemicidal activity and correlates with antibody-mediated aggregation of Treponema pallidum rare outer membrane protein. J Immunol 144:1914–1921

Blanco DR, Champion CI, Lewinski MA, Shang ES, Simkins SG, Miller JN, Lovett MA (1999) Immunization with Treponema pallidum outer membrane vesicles induces high-titer complement-dependent treponemicidal activity and aggregation of T. pallidum rare outer membrane proteins (TROMPs). J Immunol 163:2741–2746

Blom AM, Hallstrom T, Riesbeck K (2009) Complement evasion strategies of pathogens-acquisition of inhibitors and beyond. Mol Immunol 46:2808–2817

Borrow R, Aaberge IS, Santos GF, Eudey TL, Oster P et al (2005) Interlaboratory standardization of the measurement of serum bactericidal activity by using human complement against meningococcal serogroup b, strain 44/76-SL, before and after vaccination with the Norwegian MenBvac outer membrane vesicle vaccine. Clin Diagn Lab Immunol 12:970–976

Boslego J, Garcia J, Cruz C, Zollinger W, Brandt B et al (1995) Efficacy, safety, and immunogenicity of a meningococcal group B (15:P1.3) outer membrane protein vaccine in Iquique, Chile. Chilean national committee for meningococcal disease. Vaccine 13:821–829

Boutriau D, Poolman J, Borrow R, Findlow J, Domingo JD et al (2007) Immunogenicity and safety of three doses of a bivalent (B:4:p1.19,15 and B:4:p1.7-2,4) meningococcal outer membrane vesicle vaccine in healthy adolescents. Clin Vaccine Immunol 14:65–73

Bruge J, Bouveret-Le Cam N, Danve B, Rougon G, Schulz D (2004) Clinical evaluation of a group B meningococcal N-propionylated polysaccharide conjugate vaccine in adult, male volunteers. Vaccine 22:1087–1096

Camacho AI, de Souza J, Sanchez-Gomez S, Pardo-Ros M, Irache JM, Gamazo C (2011) Mucosal immunization with Shigella flexneri outer membrane vesicles induced protection in mice. Vaccine 29:8222–8229

Cartwright K, Morris R, Rumke H, Fox A, Borrow R et al (1999) Immunogenicity and reactogenicity in UK infants of a novel meningococcal vesicle vaccine containing multiple class 1 (PorA) outer membrane proteins. Vaccine 17:2612–2619

Chatterjee SN, Das J (1966) Secretory activity of Vibrio cholerae as evidenced by electron microscopy. In: Uyeda (ed) Electron Microscopy. Maruzen Co. Ltd, Tokyo

Chatterjee SN, Das J (1967) Electron microscopic observations on the excretion of cell-wall material by Vibrio cholerae. J Gen Microbiol 49:1–11

Chaturvedi A, Pierce SK (2009) How location governs toll-like receptor signaling. Traffic 10:621–628

Chaudhuri K, Chatterjee SN (2009) Cholera toxins. Springer, Heidelberg

Chen DJ, Osterrieder N, Metzger SM, Buckles E, Doody AM, DeLisa MP, Putnam D (2010) Delivery of foreign antigens by engineered outer membrane vesicle vaccines. Proc Natl Acad Sci U S A 107:3099–3104

Claassen I, Meylis J, van der Ley P, Peeters C, Brons H et al (1996) Production, characterization and control of a Neisseria meningitidis hexavalent class 1 outer membrane protein containing vesicle vaccine. Vaccine 14:1001–1008

Comanducci M, Bambini S, Brunelli B, Adu-Bobie J, Arico B et al (2002) NadA, a novel vaccine candidate of Neisseria meningitidis. J Exp Med 195:1445–1454

Cookson BT, Bevan MJ (1997) Identification of a natural T cell epitope presented by Salmonella-infected macrophages and recognized by T cells from orally immunized mice. J Immunol 158:4310–4319

Cox AD, Zou W, Gidney MA, Lacelle S, Plested JS et al (2005) Candidacy of LPS-based glycoconjugates to prevent invasive meningococcal disease: developmental chemistry and investigation of immunological responses following immunization of mice and rabbits. Vaccine 23:5045–5054

Czerkinsky C, Holmgren J (2009) Enteric vaccines for the developing world: a challenge for mucosal immunology. Mucosal Immunol 2:284–287

de Kleijn ED, de Groot R, Labadie J, Lafeber AB, van den Dobbelsteen G et al (2000) Immunogenicity and safety of a hexavalent meningococcal outer-membrane-vesicle vaccine in children of 2–3 and 7–8 years of age. Vaccine 18:1456–1466

de Moraes JC, Perkins BA, Camargo MC, Hidalgo NT, Barbosa HA et al (1992) Protective efficacy of a serogroup B meningococcal vaccine in Sao Paulo, Brazil. Lancet 340:1074–1078

del Castillo FJ, Moreno F, del Castillo I (2001) Secretion of the Escherichia coli K-12 SheA hemolysin is independent of its cytolytic activity. FEMS Microbiol Lett 204:281–285

Desvarieux M, Demmer RT, Rundek T, Boden-Albala B, Jacobs DR Jr, Sacco RL, Papapanou PN (2005) Periodontal microbiota and carotid intima-media thickness: the Oral Infections and vascular disease epidemiology study (INVEST). Circulation 111:576–582

Devoe IW, Gilchrist JE (1973) Release of endotoxin in the form of cell wall blebs during in vitro growth of Neisseria meningitidis. J Exp Med 138:1156–1167

Dharakul T, Songsivilai S (1999) The many facets of melioidosis. Trends Microbiol 7:138–140

Diacovich L, Gorvel JP (2010) Bacterial manipulation of innate immunity to promote infection. Nat Rev Microbiol 8:117–128

Dietrich G, Griot-Wenk M, Metcalfe IC, Lang AB, Viret JF (2003) Experience with registered mucosal vaccines. Vaccine 21:678–683

Drabick JJ, Brandt BL, Moran EE, Saunders NB, Shoemaker DR, Zollinger WD (1999) Safety and immunogenicity testing of an intranasal group B meningococcal native outer membrane vesicle vaccine in healthy volunteers. Vaccine 18:160–172

Durand V, Mackenzie J, de Leon J, Mesa C, Quesniaux V et al (2009) Role of lipopolysaccharide in the induction of type I interferon-dependent cross-priming and IL-10 production in mice by meningococcal outer membrane vesicles. Vaccine 27:1912–1922

Ellis TN, Kuehn MJ (2010) Virulence and immunomodulatory roles of bacterial outer membrane vesicles. Microbiol Mol Biol Rev 74:81–94

Ernst RK, Guina T, Miller SI (2001) Salmonella typhimurium outer membrane remodeling: role in resistance to host innate immunity. Microbes Infect 3:1327–1334

Estabrook MM, Jarvis GA, McLeod Griffiss J (2007) Affinity-purified human immunoglobulin G that binds a lacto-N-neotetraose-dependent lipooligosaccharide structure is bactericidal for serogroup B Neisseria meningitidis. Infect Immun 75:1025–1033

Estevez F, Carr A, Solorzano L, Valiente O, Mesa C et al (1999) Enhancement of the immune response to poorly immunogenic gangliosides after incorporation into very small size proteoliposomes (VSSP). Vaccine 18:190–197

EU-IBIS (2002) Invasive Neisseria meningitidis in Europe—2002

Faruque SM, Albert MJ, Mekalanos JJ (1998) Epidemiology, genetics, and ecology of toxigenic Vibrio cholerae. Microbiol Mol Biol Rev 62:1301–1314

Finne J, Finne U, Deagostini-Bazin H, Goridis C (1983) Occurrence of alpha 2–8 linked polysialosyl units in a neural cell adhesion molecule. Biochem Biophys Res Commun 112:482–487

Fisseha M, Chen P, Brandt B, Kijek T, Moran E, Zollinger W (2005) Characterization of native outer membrane vesicles from lpxL mutant strains of Neisseria meningitidis for use in parenteral vaccination. Infect Immun 73:4070–4080

Fransen F, Heckenberg SG, Hamstra HJ, Feller M, Boog CJ et al (2009) Naturally occurring lipid A mutants in Neisseria meningitidis from patients with invasive meningococcal disease are associated with reduced coagulopathy. PLoS Pathog 5:e1000396

Frasch CE (1995) Meningococcal vaccines: past, present and future. In: Cartwright K (ed) Meningococcal disease. Wiley, Chichester

Frasch CE, Mocca LF (1982) Strains of Neisseria meningitidis isolated from patients and their close contacts. Infect Immun 37:155–159

Frasch CE, Peppler MS (1982) Protection against group B Neisseria meningitidis disease: preparation of soluble protein and protein-polysaccharide immunogens. Infect Immun 37:271–280

Frasch CE, Robbins JD (1978) Protection against group B meningococcal disease. III. Immunogenicity of serotype 2 vaccines and specificity of protection in a guinea pig model. J Exp Med 147:629–644

Frasch CE, Coetzee G, Zahradnik JM, Feldman HA, KH J (1983) Development and evaluation of group B serotype 2 protein vaccines: report of a group B field trial. Med Trop 43:177–180

Fredriksen JH, Rosenqvist E, Wedege E, Bryn K, Bjune G et al. (1991) Production, characterization and control of MenB-vaccine "Folkehelsa": an outer membrane vesicle vaccine against group B meningococcal disease. NIPH Ann 14:67–79, discussion 79–80

Galen JE, Zhao L, Chinchilla M, Wang JY, Pasetti MF, Green J, Levine MM (2004) Adaptation of the endogenous Salmonella enterica serovar Typhi clyA-encoded hemolysin for antigen export enhances the immunogenicity of anthrax protective antigen domain 4 expressed by the attenuated live-vector vaccine strain CVD 908-htrA. Infect Immun 72:7096–7106

Garnacho J, Sole-Violan J, Sa-Borges M, Diaz E, Rello J (2003) Clinical impact of pneumonia caused by Acinetobacter baumannii in intubated patients: a matched cohort study. Crit Care Med 31:2478–2482

Geurtsen J, Vandebriel RJ, Gremmer ER, Kuipers B, Tommassen J, van der Ley P (2007) Consequences of the expression of lipopolysaccharide-modifying enzymes for the efficacy and reactogenicity of whole-cell pertussis vaccines. Microbes Infect 9:1096–1103

Giuliani MM, Santini L, Brunelli B, Biolchi A, Arico B et al (2005) The region comprising amino acids 100–255 of Neisseria meningitidis lipoprotein GNA 1870 elicits bactericidal antibodies. Infect Immun 73:1151–1160

Gold R, Goldschneider I, Lepow ML, Draper TF, Randolph M (1978) Carriage of Neisseria meningitidis and Neisseria lactamica in infants and children. J Infect Dis 137:112–121

Goldschneider I, Gotschlich EC, Artenstein MS (1969) Human immunity to the meningococcus. I. The role of humoral antibodies. J Exp Med 129:1307–1326

Gorringe AR (2005) Can Neisseria lactamica antigens provide an effective vaccine to prevent meningococcal disease? Expert Rev Vaccines 4:373–379

Gorringe A, Halliwell D, Matheson M, Reddin K, Finney M, Hudson M (2005) The development of a meningococcal disease vaccine based on Neisseria lactamica outer membrane vesicles. Vaccine 23:2210–2213

Granoff DM, Welsch JA, Ram S (2009) Binding of complement factor H (fH) to Neisseria meningitidis is specific for human fH and inhibits complement activation by rat and rabbit sera. Infect Immun 77:764–769

Guruge JL, Falk PG, Lorenz RG, Dans M, Wirth HP et al (1998) Epithelial attachment alters the outcome of Helicobacter pylori infection. Proc Natl Acad Sci U S A 95:3925–3930

Hallstrom T, Riesbeck K (2010) Hemophilus influenzae and the complement system. Trends Microbiol 18:258–265

Hallstrom T, Muller SA, Morgelin M, Mollenkvist A, Forsgren A, Riesbeck K (2008) The Moraxella IgD-binding protein MID/Hag is an oligomeric autotransporter. Microbes Infect 10:374–381

Haneberg B, Dalseg R, Oftung F, Wedege E, Hoiby EA et al (1998) Towards a nasal vaccine against meningococcal disease, and prospects for its use as a mucosal adjuvant. Dev Biol Stand 92:127–133

Haque A, Chu K, Easton A, Stevens MP, Galyov EE et al (2006) A live experimental vaccine against Burkholderia pseudomallei elicits CD4 + T cell-mediated immunity, priming T cells specific for 2 type III secretion system proteins. J Infect Dis 194:1241–1248

Healey GD, Elvin SJ, Morton M, Williamson ED (2005) Humoral and cell-mediated adaptive immune responses are required for protection against Burkholderia pseudomallei challenge and bacterial clearance postinfection. Infect Immun 73:5945–5951

Heiniger N, Spaniol V, Troller R, Vischer M, Aebi C (2007) A reservoir of Moraxella catarrhalis in human pharyngeal lymphoid tissue. J Infect Dis 196:1080–1087

Holst J (2007) Strategies for development of universal vaccines against meningococcal serogroup B disease: the most promising options and the challenges evaluating them. Hum Vaccin 3:290–294

Holst J, Feiring B, Fuglesang JE, Hoiby EA, Nokleby H, Aaberge IS, Rosenqvist E (2003) Serum bactericidal activity correlates with the vaccine efficacy of outer membrane vesicle vaccines against Neisseria meningitidis serogroup B disease. Vaccine 21:734–737

Holst J, Feiring B, Naess LM, Norheim G, Kristiansen P et al (2005) The concept of "tailor-made", protein-based, outer membrane vesicle vaccines against meningococcal disease. Vaccine 23:2202–2205

Holst J, Martin D, Arnold R, Huergo CC, Oster P, O'Hallahan J, Rosenqvist E (2009) Properties and clinical performance of vaccines containing outer membrane vesicles from Neisseria meningitidis. Vaccine 27(Suppl 2):B3–12

Hou VC, Koeberling O, Welsch JA, Granoff DM (2005) Protective antibody responses elicited by a meningococcal outer membrane vesicle vaccine with overexpressed genome-derived Neisserial antigen 1870. J Infect Dis 192:580–590

Ismail S, Hampton MB, Keenan JI (2003) Helicobacter pylori outer membrane vesicles modulate proliferation and interleukin-8 production by gastric epithelial cells. Infect Immun 71: 5670–5675

Jennings HJ, Roy R, Gamian A (1986) Induction of meningococcal group B polysaccharide-specific IgG antibodies in mice by using an N-propionylated B polysaccharide-tetanus toxoid conjugate vaccine. J Immunol 137:1708–1713

Jennings HJ, Gamian A, Ashton FE (1987) N-propionylated group B meningococcal polysaccharide mimics a unique epitope on group B Neisseria meningitidis. J Exp Med 165: 1207–1211

Jennings GT, Savino S, Marchetti E, Arico B, Kast T et al (2002) GNA33 from Neisseria meningitidis serogroup B encodes a membrane-bound lytic transglycosylase (MltA). Eur J Biochem 269:3722–3731

Jin JS, Kwon SO, Moon DC, Gurung M, Lee JH, Kim SI, Lee JC (2011) Acinetobacter baumannii secretes cytotoxic outer membrane protein A via outer membrane vesicles. PLoS ONE 6:e17027

John TJ, Jesudason MV, Lalitha MK, Ganesh A, Mohandas V et al (1996) Melioidosis in India: the tip of the iceberg? Indian J Med Res 103:62–65

Jones DM, Eldridge J (1979) Development of antibodies to meningococcal protein and lipopolysaccharide serotype antigens in healthy-carriers. J Med Microbiol 12:107–111

Jones BD, Falkow S (1996) Salmonellosis: host immune responses and bacterial virulence determinants. Annu Rev Immunol 14:533–561

Kaminski RW, Oaks EV (2009) Inactivated and subunit vaccines to prevent shigellosis. Expert Rev Vaccines 8:1693–1704

Kaparakis M, Turnbull L, Carneiro L, Firth S, Coleman HA et al (2010) Bacterial membrane vesicles deliver peptidoglycan to NOD1 in epithelial cells. Cell Microbiol 12:372–385

Kaper JB, Morris JG Jr, Levine MM (1995) Cholera. Clin Microbiol Rev 8:48–86

Katial RK, Brandt BL, Moran EE, Marks S, Agnello V, Zollinger WD (2002) Immunogenicity and safety testing of a group B intranasal meningococcal native outer membrane vesicle vaccine. Infect Immun 70:702–707

Kawai T, Akira S (2010) The role of pattern-recognition receptors in innate immunity: update on toll-like receptors. Nat Immunol 11:373–384

Keenan JI, Allardyce RA, Bagshaw PF (1998) Lack of protection following immunisation with H. pylori outer membrane vesicles highlights antigenic differences between H. felis and H. pylori. FEMS Microbiol Lett 161:21–27

Keenan JI, Rijpkema SG, Durrani Z, Roake JA (2003) Differences in immunogenicity and protection in mice and guinea pigs following intranasal immunization with Helicobacter pylori outer membrane antigens. FEMS Immunol Med Microbiol 36:199–205

Kelly DF, Rappuoli R (2005) Reverse vaccinology and vaccines for serogroup B Neisseria meningitidis. Adv Exp Med Biol 568:217–223

Kesavalu L, Ebersole JL, Machen RL, Holt SC (1992) Porphyromonas gingivalis virulence in mice: induction of immunity to bacterial components. Infect Immun 60:1455–1464

Kim JY, Doody AM, Chen DJ, Cremona GH, Shuler ML, Putnam D, DeLisa MP (2008) Engineered bacterial outer membrane vesicles with enhanced functionality. J Mol Biol 380:51–66

Koeberling O, Seubert A, Granoff DM (2008) Bactericidal antibody responses elicited by a meningococcal outer membrane vesicle vaccine with overexpressed factor H-binding protein and genetically attenuated endotoxin. J Infect Dis 198:262–270

Koeberling O, Giuntini S, Seubert A, Granoff DM (2009) Meningococcal outer membrane vesicle vaccines derived from mutant strains engineered to express factor H binding proteins from antigenic variant groups 1 and 2. Clin Vaccine Immunol 16:156–162

Koser ML, McGettigan JP, Tan GS, Smith ME, Koprowski H, Dietzschold B, Schnell MJ (2004) Rabies virus nucleoprotein as a carrier for foreign antigens. Proc Natl Acad Sci U S A 101:9405–9410

Kotloff KL, Winickoff JP, Ivanoff B, Clemens JD, Swerdlow DL et al (1999) Global burden of shigella infections: implications for vaccine development and implementation of control strategies. Bull World Health Organ 77:651–666

Kouokam JC, Wai SN (2006) Outer membrane vesicle-mediated export of a poreforming cytotoxin from Escherichia coli. Toxin Rev 25:31–46

Kuehn MJ, Kesty NC (2005) Bacterial outer membrane vesicles and the host-pathogen interaction. Genes Dev 19:2645–2655

Kulshin VA, Zahringer U, Lindner B, Frasch CE, Tsai CM, Dmitriev BA, Rietschel ET (1992) Structural characterization of the lipid A component of pathogenic Neisseria meningitidis. J Bacteriol 174:1793–1800

Kurosaki T, Shinohara H, Baba Y (2010) B cell signaling and fate decision. Annu Rev Immunol 28:21–55

Kweon MN (2008) Shigellosis: the current status of vaccine development. Curr Opin Infect Dis 21:313–318

Kwon SO, Gho YS, Lee JC, Kim SI (2009) Proteome analysis of outer membrane vesicles from a clinical Acinetobacter baumannii isolate. FEMS Microbiol Lett 297:150–156

Lamont RJ, Jenkinson HF (1998) Life below the gum line: pathogenic mechanisms of Porphyromonas gingivalis. Microbiol Mol Biol Rev 62:1244–1263

Lee SR, Kim SH, Jeong KJ, Kim KS, Kim YH et al (2009) Multi-immunogenic outer membrane vesicles derived from an MsbB-deficient Salmonella enterica serovar typhimurium mutant. J Microbiol Biotechnol 19:1271–1279

Levine MM, Kotloff KL, Barry EM, Pasetti MF, Sztein MB (2007) Clinical trials of Shigella vaccines: two steps forward and one step back on a long, hard road. Nat Rev Microbiol 5:540–553

Locht C (2008) A common vaccination strategy to solve unsolved problems of tuberculosis and pertussis? Microbes Infect 10:1051–1056

Low KB, Ittensohn M, Le T, Platt J, Sodi S et al (1999) Lipid A mutant salmonella with suppressed virulence and TNFalpha induction retain tumor-targeting in vivo. Nat Biotechnol 17:37–41

Lowell GH, Ballou WR, Smith LF, Wirtz RA, Zollinger WD, Hockmeyer WT (1988) Proteosome-lipopeptide vaccines: enhancement of immunogenicity for malaria CS peptides. Science 240:800–802

Lowrie DB, Aber VR, Carrol ME (1979) Division and death rates of salmonella typhimurium inside macrophages: use of penicillin as a probe. J Gen Microbiol 110:409–419

Magalhaes JG, Philpott DJ, Nahori MA, Jehanno M, Fritz J et al (2005) Murine nod1 but not its human orthologue mediates innate immune detection of tracheal cytotoxin. EMBO Rep 6:1201–1207

Mallett CP, Hale TL, Kaminski RW, Larsen T, Orr N, Cohen D, Lowell GH (1995) Intransal or intragastric immunization with proteosome-shigella lipopolysaccharide vaccines protects against lethal pneumonia in a murine model of shigella infection. Infect Immun 63:2382–2386

Martin D, McDowell R (2004) The epidemiology of meningococcal disease in New Zealand in 2003. Ministry of health by the institute of environmental science and research (ESR) Limited, Wellington

Martin DR, Walker SJ, Baker MG, Lennon DR (1998) New Zealand epidemic of meningococcal disease identified by a strain with phenotype B:4:P1.4. J Infect Dis 177:497–500

Masignani V, Balducci E, Di Marcello F, Savino S, Serruto D et al (2003a) NarE: a novel ADP-ribosyltransferase from Neisseria meningitidis. Mol Microbiol 50:1055–1067

Masignani V, Comanducci M, Giuliani MM, Bambini S, Adu-Bobie J et al (2003b) Vaccination against Neisseria meningitidis using three variants of the lipoprotein GNA1870. J Exp Med 197:789–799

Mastroeni P, Villarreal-Ramos B, Hormaeche CE (1993) Adoptive transfer of immunity to oral challenge with virulent salmonellae in innately susceptible BALB/c mice requires both immune serum and T cells. Infect Immun 61:3981–3984

McConnell MJ, Pachon J (2010) Active and passive immunization against Acinetobacter baumannii using an inactivated whole cell vaccine. Vaccine 29:1–5

McConnell MJ, Rumbo C, Bou G, Pachon J (2011) Outer membrane vesicles as an acellular vaccine against Acinetobacter baumannii. Vaccine 29:5705–5710

McSorley SJ, Cookson BT, Jenkins MK (2000) Characterization of CD4 + T cell responses during natural infection with Salmonella typhimurium. J Immunol 164:986–993

Mitka M (2005) New vaccine should ease meningitis fears. JAMA 293:1433–1434

Moe GR, Zuno-Mitchell P, Hammond SN, Granoff DM (2002) Sequential immunization with vesicles prepared from heterologous Neisseria meningitidis strains elicits broadly protective serum antibodies to group B strains. Infect Immun 70:6021–6031

Morein B, Sundquist B, Hoglund S, Dalsgaard K, Osterhaus A (1984) Iscom, a novel structure for antigenic presentation of membrane proteins from enveloped viruses. Nature 308:457–460

Moreno C, Lifely MR, Esdaile J (1985) Immunity and protection of mice against Neisseria meningitidis group B by vaccination, using polysaccharide complexed with outer membrane proteins: a comparison with purified B polysaccharide. Infect Immun 47:527–533

Nakao R, Hasegawa H, Ochiai K, Takashiba S, Ainai A et al (2011) Outer membrane vesicles of Porphyromonas gingivalis elicit a mucosal immune response. PLoS ONE 6:e26163

Nieves W, Asakrah S, Qazi O, Brown KA, Kurtz J et al (2011) A naturally derived outer-membrane vesicle vaccine protects against lethal pulmonary Burkholderia pseudomallei infection. Vaccine 29:8381–8389

Nolan T, Lambert S, Roberton D, Marshall H, Richmond P et al (2007) A novel combined hemophilus influenzae type b-Neisseria meningitidis serogroups C and Y-tetanus-toxoid conjugate vaccine is immunogenic and induces immune memory when co-administered with DTPa-HBV-IPV and conjugate pneumococcal vaccines in infants. Vaccine 25:8487–8499

Nordstrom T, Forsgren A, Riesbeck K (2002) The immunoglobulin D-binding part of the outer membrane protein MID from Moraxella catarrhalis comprises 238 amino acids and a tetrameric structure. J Biol Chem 277:34692–34699

Noronha CP, Struchiner CJ, Halloran ME (1995) Assessment of the direct effectiveness of BC meningococcal vaccine in Rio de Janeiro, Brazil: a case-control study. Int J Epidemiol 24:1050–1057

Ntezayabo B, De Serres G, Duval B (2003) Pertussis resurgence in Canada largely caused by a cohort effect. Pediatr Infect Dis J 22:22–27

O'Hagan DT, Valiante NM (2003) Recent advances in the discovery and delivery of vaccine adjuvants. Nat Rev Drug Discov 2:727–735

Oliver KJ, Reddin KM, Bracegirdle P, Hudson MJ, Borrow R et al (2002) Neisseria lactamica protects against experimental meningococcal infection. Infect Immun 70:3621–3626

Oster P, Lennon D, O'Hallahan J, Mulholland K, Reid S, Martin D (2005) MeNZB: a safe and highly immunogenic tailor-made vaccine against the New Zealand Neisseria meningitidis serogroup B disease epidemic strain. Vaccine 23:2191–2196

Oster P, O'Hallahan J, Aaberge I, Tilman S, Ypma E, Martin D (2007) Immunogenicity and safety of a strain-specific MenB OMV vaccine delivered to under 5-year olds in New Zealand. Vaccine 25:3075–3079

Park SB, Jang HB, Nho SW, Cha IS, Hikima J et al (2011) Outer membrane vesicles as a candidate vaccine against edwardsiellosis. PLoS ONE 6:e17629

Parkhill J, Achtman M, James KD, Bentley SD, Churcher C et al (2000) Complete DNA sequence of a serogroup A strain of Neisseria meningitidis Z2491. Nature 404:502–506

Patial S, Chaturvedi VK, Rai A, Saini M, Chandra R, Saini Y, Gupta PK (2007) Virus neutralizing antibody response in mice and dogs with a bicistronic DNA vaccine encoding rabies virus glycoprotein and canine parvovirus VP2. Vaccine 25:4020–4028

Perkins BA, Jonsdottir K, Briem H, Griffiths E, Plikaytis BD et al (1998) Immunogenicity of two efficacious outer membrane protein-based serogroup B meningococcal vaccines among young adults in Iceland. J Infect Dis 177:683–691

Perrett KP, Pollard AJ (2005) Towards an improved serogroup B Neisseria meningitidis vaccine. Expert Opin Biol Ther 5:1611–1625

Pierson T, Matrakas D, Taylor YU, Manyam G, Morozov VN, Zhou W, van Hoek ML (2011) Proteomic characterization and functional analysis of outer membrane vesicles of Francisella novicida suggests possible role in virulence and use as a vaccine. J Proteome Res 10:954–967

Pizza M, Scarlato V, Masignani V, Giuliani MM, Arico B et al (2000) Identification of vaccine candidates against serogroup B meningococcus by whole-genome sequencing. Science 287:1816–1820

Pizza M, Giuliani MM, Fontana MR, Monaci E, Douce G et al (2001) Mucosal vaccines: non toxic derivatives of LT and CT as mucosal adjuvants. Vaccine 19:2534–2541

Plested JS, Harris SL, Wright JC, Coull PA, Makepeace K et al (2003) Highly conserved Neisseria meningitidis inner-core lipopolysaccharide epitope confers protection against experimental meningococcal bacteremia. J Infect Dis 187:1223–1234

Roberts R, Moreno G, Bottero D, Gaillard ME, Fingermann M et al (2008) Outer membrane vesicles as acellular vaccine against pertussis. Vaccine 26:4639–4646

Rodriguez AP, Dickinson F, Baly A, Martinez R (1999) The epidemiological impact of antimeningococcal B vaccination in Cuba. Mem Inst Oswaldo Cruz 94:433–440

Rosenstein NE, Perkins BA, Stephens DS, Popovic T, Hughes JM (2001) Meningococcal disease. N Engl J Med 344:1378–1388

Roy N, Barman S, Ghosh A, Pal A, Chakraborty K et al (2010) Immunogenicity and protective efficacy of Vibrio cholerae outer membrane vesicles in rabbit model. FEMS Immunol Med Microbiol 60:18–27

Roy K, Hamilton DJ, Munson GP, Fleckenstein JM (2011) Outer membrane vesicles induce immune responses to virulence proteins and protect against colonization by enterotoxigenic Escherichia coli. Clin Vaccine Immunol 18:1803–1808

Sack DA, Sack RB, Nair GB, Siddique AK (2004) Cholera Lancet 363:223–233

Sadarangani M, Pollard AJ (2010) Serogroup B meningococcal vaccines-an unfinished story. Lancet Infect Dis 10:112–124

Samuelsson M, Jendholm J, Amisten S, Morrison SL, Forsgren A, Riesbeck K (2006) The IgD CH1 region contains the binding site for the human respiratory pathogen Moraxella catarrhalis IgD-binding protein MID. Eur J Immunol 36:2525–2534

Santos GF, Deck RR, Donnelly J, Blackwelder W, Granoff DM (2001) Importance of complement source in measuring meningococcal bactericidal titers. Clin Diagn Lab Immunol 8:616–623

Sardinas G, Reddin K, Pajon R, Gorringe A (2006) Outer membrane vesicles of Neisseria lactamica as a potential mucosal adjuvant. Vaccine 24:206–214

Sarkar-Tyson M, Titball RW (2010) Progress toward development of vaccines against melioidosis: a review. Clin Ther 32:1437–1445

Saunders NB, Shoemaker DR, Brandt BL, Moran EE, Larsen T, Zollinger WD (1999) Immunogenicity of intranasally administered meningococcal native outer membrane vesicles in mice. Infect Immun 67:113–119

Schild S, Nelson EJ, Camilli A (2008) Immunization with vibrio cholerae outer membrane vesicles induces protective immunity in mice. Infect Immun 76:4554–4563

Schild S, Nelson EJ, Bishop AL, Camilli A (2009) Characterization of vibrio cholerae outer membrane vesicles as a candidate vaccine for cholera. Infect Immun 77:472–484

Serruto D, Adu-Bobie J, Scarselli M, Veggi D, Pizza M, Rappuoli R, Arico B (2003) Neisseria meningitidis app, a new adhesin with autocatalytic serine protease activity. Mol Microbiol 48:323–334

Serruto D, Spadafina T, Ciucchi L, Lewis LA, Ram S et al (2010) Neisseria meningitidis GNA2132, a heparin-binding protein that induces protective immunity in humans. Proc Natl Acad Sci U S A 107:3770–3775

Sethi S, Sethi R, Eschberger K, Lobbins P, Cai X, Grant BJ, Murphy TF (2007) Airway bacterial concentrations and exacerbations of chronic obstructive pulmonary disease. Am J Respir Crit Care Med 176:356–361

Shang ES, Champion CI, Wu XY, Skare JT, Blanco DR, Miller JN, Lovett MA (2000) Comparison of protection in rabbits against host-adapted and cultivated Borrelia burgdorferi following infection-derived immunity or immunization with outer membrane vesicles or outer surface protein A. Infect Immun 68:4189–4199

Sharifat Salmani A, Siadat SD, Norouzian D, Izadi Mobarakeh J, Kheirandish M, Zangeneh M (2009) Outer membrane vesicle of Neisseria meningitidis serogroup B as an adjuvant to induce specific antibody response against the lipopolysaccharide of Brucella abortus S99. Ann Microbiol 1:145–149

Sharip A, Sorvillo F, Redelings MD, Mascola L, Wise M, Nguyen DM (2006) Population-based analysis of meningococcal disease mortality in the United States: 1990–2002. Pediatr Infect Dis J 25:191–194

Shepard CW, Rosenstein NE, Fischer M (2003) Neonatal meningococcal disease in the United States, 1990–1999. Pediatr Infect Dis J 22:418–422

Siadat SD, Kheirandish M, Norouzian D, Behzadiyannejad Q, Najar Peerayeh S, Zangeneh M, Nejati M (2007) A flow cytometric opsonophagocytic assay for measurement of functional antibodies elicited after immunization with outer membrane vesicle of Neisseria meningitidis serogroup B. Pak J Biol Sci 10:3578–3584

Siadat SD, Naddaf SR, Zangeneh M, Moshiri A, Sadat SM et al (2011) Outer membrane vesicle of Neisseria meningitidis serogroup B as an adjuvant in immunization of rabbit against Neisseria meningitidis serogroup A. Afr J Microbiol Res 5:3090–3095

Sierra GV, Campa HC, Varcacel NM, Garcia IL, Izquierdo PL et al (1991) Vaccine against group B Neisseria meningitidis: protection trial and mass vaccination results in Cuba. NIPH Ann 14:195–207, discussion 208–110

Singh M, Chakrapani A, O'Hagan D (2007) Nanoparticles and microparticles as vaccine-delivery systems. Expert Rev Vaccines 6:797–808

Singh B, Su YC, Riesbeck K (2010) Vitronectin in bacterial pathogenesis: a host protein used in complement escape and cellular invasion. Mol Microbiol 78:545–560

Singh K, Bayrak B, Riesbeck K (2012) A role for TLRs in Moraxella-superantigen induced polyclonal B cell activation. Front Biosci (Schol Ed) 4:1031–1043

Steeghs L, Keestra AM, van Mourik A, Uronen-Hansson H, van der Ley P et al (2008) Differential activation of human and mouse toll-like receptor 4 by the adjuvant candidate LpxL1 of Neisseria meningitidis. Infect Immun 76:3801–3807

Stephens DS, Greenwood B, Brandtzaeg P (2007) Epidemic meningitis, meningococcaemia, and Neisseria meningitidis. Lancet 369:2196–2210

Su EL, Snape MD (2011) A combination recombinant protein and outer membrane vesicle vaccine against serogroup B meningococcal disease. Expert Rev Vaccines 10:575–588

Sztein MB, Wasserman SS, Tacket CO, Edelman R, Hone D, Lindberg AA, Levine MM (1994) Cytokine production patterns and lymphoproliferative responses in volunteers orally immunized with attenuated vaccine strains of Salmonella typhi. J Infect Dis 170:1508–1517

Tan TT, Nordstrom T, Forsgren A, Riesbeck K (2005) The respiratory pathogen Moraxella catarrhalis adheres to epithelial cells by interacting with fibronectin through ubiquitous surface proteins A1 and A2. J Infect Dis 192:1029–1038

Tan TT, Morgelin M, Forsgren A, Riesbeck K (2007) Hemophilus influenzae survival during complement-mediated attacks is promoted by Moraxella catarrhalis outer membrane vesicles. J Infect Dis 195:1661–1670

Tanaka M, Vitek CR, Pascual FB, Bisgard KM, Tate JE, Murphy TV (2003) Trends in pertussis among infants in the United States, 1980–1999. JAMA 290:2968–2975

Tappero JW, Lagos R, Ballesteros AM, Plikaytis B, Williams D et al (1999) Immunogenicity of 2 serogroup B outer-membrane protein meningococcal vaccines: a randomized controlled trial in Chile. JAMA 281:1520–1527

Tettelin H, Saunders NJ, Heidelberg J, Jeffries AC, Nelson KE et al (2000) Complete genome sequence of Neisseria meningitidis serogroup B strain MC58. Science 287:1809–1815

Tondella ML, Popovic T, Rosenstein NE, Lake DB, Carlone GM, Mayer LW, Perkins BA (2000) Distribution of Neisseria meningitidis serogroup B serosubtypes and serotypes circulating in the United States. The active bacterial core surveillance team. J Clin Microbiol 38:3323–3328

Trotter CL, Ramsay ME (2007) Vaccination against meningococcal disease in Europe: review and recommendations for the use of conjugate vaccines. FEMS Microbiol Rev 31:101–107

Trotter C, Findlow J, Balmer P, Holland A, Barchha R et al (2007) Seroprevalence of bactericidal and anti-outer membrane vesicle antibodies to Neisseria meningitidis group B in England. Clin Vaccine Immunol 14:863–868

Urwin R, Russell JE, Thompson EA, Holmes EC, Feavers IM, Maiden MC (2004) Distribution of surface protein variants among hyperinvasive meningococci: implications for vaccine design. Infect Immun 72:5955–5962

van Deuren M, Brandtzaeg P, van der Meer JW (2000) Update on meningococcal disease with emphasis on pathogenesis and clinical management. Clin Microbiol Rev 13:144–166, table of contents

Vaughan AT, Brackenbury LS, Massari P, Davenport V, Gorringe A, Heyderman RS, Williams NA (2010) Neisseria lactamica selectively induces mitogenic proliferation of the naive B cell pool via cell surface Ig. J Immunol 185:3652–3660

Viala J, Chaput C, Boneca IG, Cardona A, Girardin SE et al (2004) Nod1 responds to peptidoglycan delivered by the Helicobacter pylori cag pathogenicity island. Nat Immunol 5:1166–1174

Vidakovics ML, Jendholm J, Morgelin M, Mansson A, Larsson C, Cardell LO, Riesbeck K (2010) B cell activation by outer membrane vesicles–a novel virulence mechanism. PLoS Pathog 6:e1000724

Vogel H, Jahnig F (1986) Models for the structure of outer-membrane proteins of Escherichia coli derived from raman spectroscopy and prediction methods. J Mol Biol 190:191–199

Wai SN, Lindmark B, Soderblom T, Takade A, Westermark M et al (2003) Vesicle-mediated export and assembly of pore-forming oligomers of the enterobacterial ClyA cytotoxin. Cell 115:25–35

Walker EM, Zampighi GA, Blanco DR, Miller JN, Lovett MA (1989) Demonstration of rare protein in the outer membrane of Treponema pallidum subsp. pallidum by freeze-fracture analysis. J Bacteriol 171:5005–5011

Wang L, Huang JA, Nagesha HS, Smith SC, Phelps A et al (1999) Bacterial expression of the major antigenic regions of porcine rotavirus VP7 induces a neutralizing immune response in mice. Vaccine 17:2636–2645

Wedege E, Nokleby H, Bjune G (1999) No evidence for serosubtype-restricted protection among teenagers vaccinated with the Norwegian group B outer membrane vesicle vaccine. J Infect Dis 180:242, author reply 242–243

Weynants V, Denoel P, Devos N, Janssens D, Feron C et al (2009) Genetically modified L3,7 and L2 lipooligosaccharides from Neisseria meningitidis serogroup B confer a broad cross-bactericidal response. Infect Immun 77:2084–2093

White NJ (2003) Melioidosis Lancet 361:1715–1722

WHO (1998) Control of epidemic meninogococcal disease: WHO practical guidelines. World Health organization, Geneva

Wu Y, Przysiecki C, Flanagan E, Bello-Irizarry SN, Ionescu R et al (2006) Sustained high-titer antibody responses induced by conjugating a malarial vaccine candidate to outer-membrane protein complex. Proc Natl Acad Sci U S A 103:18243–18248

Wyle FA, Artenstein MS, Brandt BL, Tramont EC, Kasper DL et al (1972) Immunologic response of man to group B meningococcal polysaccharide vaccines. J Infect Dis 126:514–521

Zimmer SM, Stephens DS (2006) Serogroup B meningococcal vaccines. Curr Opin Investig Drugs 7:733–739

Zipfel PF, Skerka C (2009) Complement regulators and inhibitory proteins. Nat Rev Immunol 9:729–740

Zollinger WD, Mandrell RE (1983) Importance of complement source in bactericidal activity of human antibody and murine monoclonal antibody to meningococcal group B polysaccharide. Infect Immun 40:257–264

Zollinger WD, Kasper DL, Veltri BJ, Artenstein MS (1972) Isolation and characterization of a native cell wall complex from Neisseria meningitidis. Infect Immun 6:835–851

Zollinger WD, Mandrell RE, Altieri P, Berman S, Lowenthal J, Artenstein MS (1978) Safety and immunogenicity of a Neisseria meningitidis type 2 protein vaccine in animals and humans. J Infect Dis 137:728–739

Zollinger WD, Donets M, Brandt BL et al (2008) Multivalent group B meningococcal vaccine based on native outer membrane vesicles has potential for providing safe, broadly protective immunity. Abstract 035.

Zollinger WD, Donets MA, Schmiel DH, Pinto VB, Labrie JE 3rd et al (2010) Design and evaluation in mice of a broadly protective meningococcal group B native outer membrane vesicle vaccine. Vaccine 28:5057–5067

Chapter 10
Concluding Notes

Abstract While pointing out some of the salient features of the up-to-date research on outer membrane vesicles (OMVs) and the unsolved or unanswered questions arising there from, attempts have been made to figure out the directions in which research on OMVs is expected to proceed in future.

Keywords OMV production · Genetics · Energetics · Signaling mechanism · Sorting process · Biofilm · Proteomic analyses · Engineered recombinant OMVs · Adjuvants · Immunogens · Vaccines

Ever since their discovery in 1966–1967, research on OMVs has progressed steadily and many more investigators have contributed significantly towards revealing the structure, both physical and chemical, and functions of the OMVs derived from numerous Gram-negative bacteria. But the search for knowledge is an unending process and accordingly, there are many more features of the OMVs that have remained unexplored or unanswered until now. In this context, the present authors have taken the opportunity or liberty to do some wild thinking, which is hoped to be relevant, about the past, present and future of the OMVs and their applications towards the welfare of mankind. Some attempts have also been made to offer questions, right or wrong, that may stimulate the minds of future investigators.While considering the different facets of these nanoparticles, this monograph has dealt with (1) bacterial growth conditions, (2) pathogenicity, (3) impact of LPS structure, (4) antibiotic treatment, (5) stress response, and so on as conditions affecting the production of OMVs by Gram-negative bacteria in general. But a detailed study of the genetics of the organisms controlling the OMV production under any situation still remains to be worked out, although some isolated studies have already been reported. Again not enough information is available on the genes, in particular, exerting positive or negative regulation on the production of OMVs. Why the OMVs are produced under such diverse conditions

S. N. Chatterjee and K. Chaudhuri, *Outer Membrane Vesicles of Bacteria*,
SpringerBriefs in Microbiology, DOI: 10.1007/978-3-642-30526-9_10,
© The Author(s) 2012

and what are the driving forces working under those conditions are some of the disturbing questions that remain to be answered.

The production of OMVs has been interpreted as a novel secretion mechanism of Gram-negative bacteria. Although there are several different secretion mechanisms operating in Gram-negative bacteria, the reason or the signal for evolving this additional secretion mechanism, production of OMVs, has remained unknown. Also the details of the energetics of this mechanism remain to be worked out.It has now been recognized that different virulence factors, proteins, toxins, antibiotics, genetic elements, and the like are packaged into OMVs for transfer and delivery into different prokaryotic and eukaryotic cells as hosts. There are enough reasons to believe that such transfer through OMVs ensures that the packaged materials are not damaged or destroyed by the hostile environment in the extra-cellular space. It is implied that some signaling mechanism works between the bacterial cell and the outer medium so as to ensure that the materials to be secreted do not face any hostile environment outside. But what this signaling mechanism and, if present at all, how it works are matters that deserve the attention of future investigators. It is known that a signaling molecule, PQS, works in *Pseudomonas aeruginosa* to control in some way or other the production of OMVs, but, here again, the exact molecular mechanism involved and particularly its genetic basis remains to be elucidated.

Investigations have revealed that not all the proteins of the outer membrane (OM) or of the periplasm (PPM) are transferred and packaged into OMVs in equal proportions and that a selection process works and controls this affair. There are reasons to believe that a sorting process works at this stage. But what is the nature of this sorting mechanism and what is its genetic basis are not known at present.

Formation of biofilm is a defense mechanism of bacteria and particularly the pathogenic ones. Enough evidence has been presented about the release of OMVs by the bacteria in the biofilm state and plenty of OMVs have been found in association with the biofilm. Whether the OMVs contribute towards formation of the biofilm or they simply help the survival of the parent bacteria in the biofilm or both is an intriguing question.

Recent proteomic analyses of the Gram-negative bacterial OMVs using mass spectrometry, ultracentrifugation, gel electrophoresis, and the like have opened a new chapter in OMV research, but have already produced results that are debated. The presence of cytoplasmic and inner membrane proteins in the OMVs is difficult to explain. This demands a very stringent method of isolation and purification of OMVs. Only a synchronized cell culture can ensure to a great extent the absence of some bacteria undergoing lysis during the logarithmic phase of growth. Such lytic bacteria may release the cytoplasmic and inner membrane proteins in the extra-cellular medium to be picked up by the OMVs. Also a very stringent method of isolation and purification of OMVs is required to ensure the absence of any contamination from the culture filtrate. Nevertheless, the analyses of proteomic profiles of OMVs derived from different Gram-negative bacteria are very impor-tant not only for elucidating the functional aspects of the different vesicular proteins but also for designing the proper mode of delivery of antigens and toxins for production of effective vaccines against Gram-negative pathogens.

Many biological molecules, proteins, toxins, antibiotics, and so on cannot directly enter different eukaryotic cells, but have found access into these cells through the OMVs. This is an important and biologically useful function of the OMVs. However, the packaging of all such molecules into the OMVs from the parent bacterial cells may not be easy or at all possible in as much as all of them cannot simply pass through or get translocated across the inner membrane to get into the PPM. In this respect, the autolysin molecule, ClyA, has the unique property of easily getting translocated across the inner membrane to reach the PPM and then to the OMVs. Not only that, ClyA can combine with different molecules and this combination can also easily get translocated across the inner membrane and entrapped into the OMVs. This gives a unique scope for producing the so-called engineered recombinant OMVs and to broaden thereby the functional spectrum of the OMVs. These engineered recombinant OMVs will, in future, play a great role in modulating the immunogenicity of the eukaryotic systems and in producing effective vaccines against many Gram-negative bacterial pathogens.

The OMV vaccines are acellular and regarded as safe for use in humans and should thus be preferred to conventional live or heat-killed or virulence-attenuated vaccines. The advantages of OMVs over subunit vaccines such as purified proteins are (1) purification of OMVs need simple ultracentifugation, effectively elimi-nating the requirement of costly infrastructure; (2) OMVs are effective as vehicles for vaccine delivery so that adjuvants are not required and (3) improvement of efficacy and safety can be done by addition of effective antigens and/or detoxifi-cation of LPS through simple genetic engineering of bacterial strains used for OMV production. The development of OMV-vaccines against different bacterial pathogens thus has a bright future.

The concept of "OMV vaccines" was developed during the 1980s when it was tried against group B meningococcal diseases. These vaccines have a long-standing safety and effectiveness record as available through a number of clinical trials. The collaborative study of the New Zealand ministry of health (NIPH) and the private firm, Chiron/Novartis, revealed that OMV vaccines can be upscaled and the quality, consistency and cost-effectiveness can be maintained. This gives the scope and direction in which further studies can be made towards improving the efficacy and quality of the vaccine.

Recent developments including sequencing of complete genomes and the reverse vaccinology approach have enabled us to predict a number of protective antigens. Technologies are at present available for determining the cross-protec-tive and universal antigens to be included in the future OMV vaccine. Thus in the coming few years, a global OMV-based vaccine strategy incorporating suitable antigens is expected to be developed and might offer protection to susceptible individuals against diseases caused by the *meningococci* of all serogroups.

After the first report of the OMV vaccines against the serogoup B meningo-coccal disease, OMVs from various other organisms including *Vibrio cholerae*, *Salmonella typhimurium*, *P. aeruginosa*, *Bordetella pertussis*, and others were shown to exhibit immunogenic properties and were reported to have vaccine potential. The future will thus see a lot of research and clinical efforts towards development and use of effective immunogens and/or vaccines against many Gram-negative organisms, in addition to *N.meningitidis*. Thus, the OMVs have already crossed the boundaries of the research laboratory and reached the open field, where the researchers and the clinicians will have a lot of scope, in future, to interact with each other for the welfare of the living world.

Index